Plastics

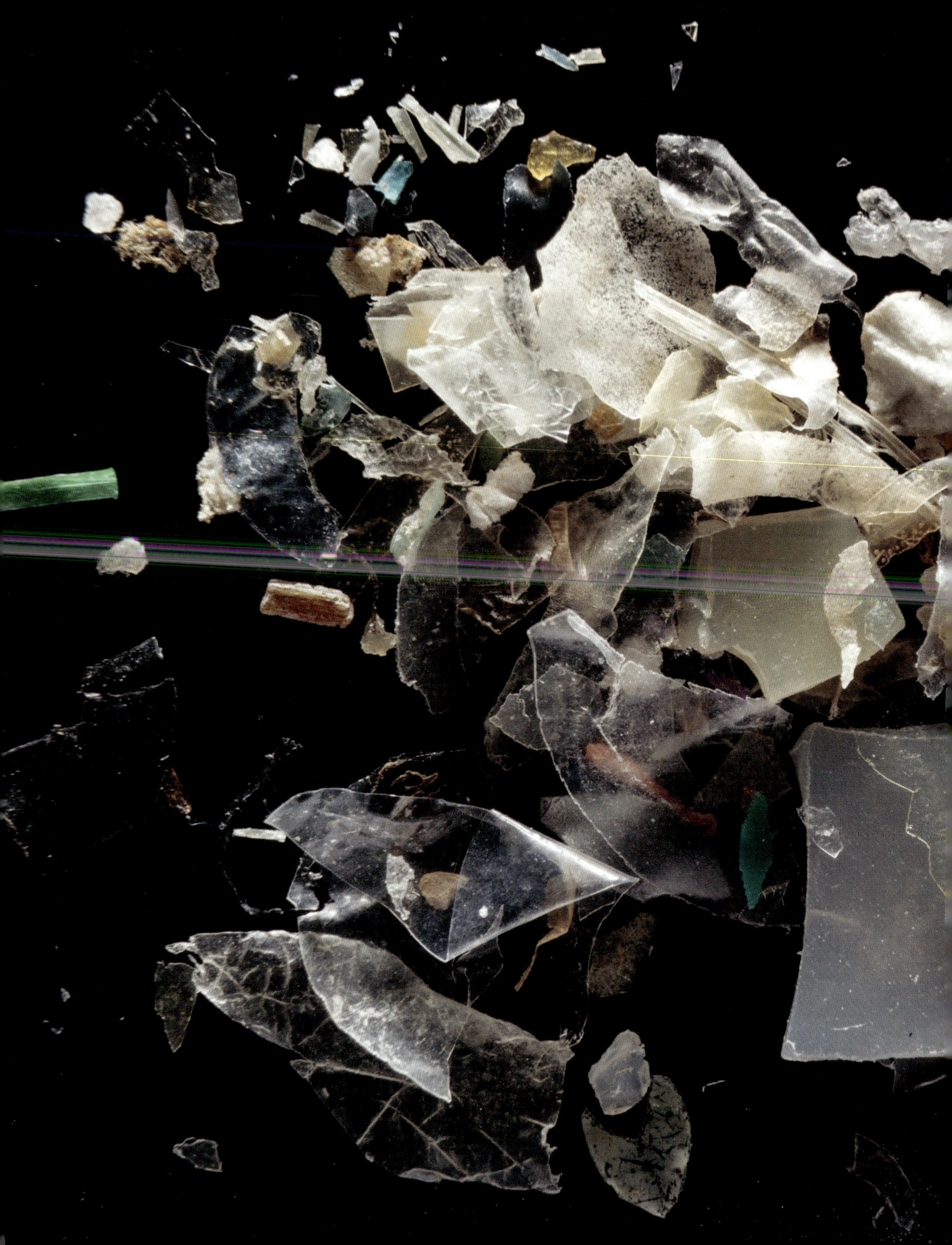

ART/WORK Edited by Caroline Fowler and Ittai Weinryb

Plastics

Anne Gunnison
David Joselit

Princeton University Press
Princeton and Oxford

ISBN (pbk.) 978-0-691-23882-1
ISBN (ebook) 978-0-691-27435-5

Library of Congress Control Number: 2025936041

British Library Cataloging-in-Publication Data is available

Published by
Princeton University Press
41 William Street
Princeton, New Jersey 08540

In the United Kingdom:
Princeton University Press
99 Banbury Road
Oxford OX2 6JX

press.princeton.edu

Editorial:
Michelle Komie and Annie Miller

Production Editorial:
Terri O'Prey

Design and composition:
Binocular, New York

Production: Steven Sears

Publicity: Jodi Price

Copyeditor: Cindy Milstein

This book has been composed in Sole Serif and Sole Sans.

GPSR Authorized Representative:
Easy Access System Europe
Mustamäe tee 50
10621 Tallinn, Estonia
gpsr.requests@easproject.com

Cover images (Front, left to right): Eva Hesse, *Repetition Nineteen* (detail). Digital image © The Museum of Modern Art / Licensed by SCALA / Art Resource, NY. © The Estate of Eva Hesse. Courtesy Hauser & Wirth. Naum Gabo, *Linear Construction No. 2* (detail), 1949–52. Presented by the artist through the American Federation of Arts, 1969. © Nina & Graham Williams / Tate, 2020. Photo: © Tate. Photograph of microplastics sample (detail). Chesapeake Bay Program. Photo: Will Parson. (Back, left to right): Photograph of deteriorated nitrate motion picture film decomposed in powder form (detail). National Archives, NAID 12168106. Kuskwogmiut Yup'ik, *Bag* (detail), ca. 1900. National Museum of the American Indian, Washington, D.C.

Frontispiece: Sample of microplastics collected from four tidal tributaries to the northern Chesapeake Bay (detail of fig. 33).

Printed in China
10 9 8 7 6 5 4 3 2 1

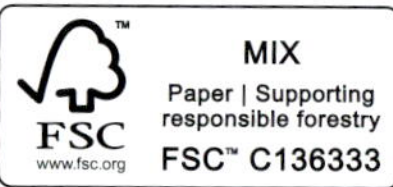

Contents

Plastics

Plastics: An Introduction

Caroline Fowler

There are infinite origin stories for plastics. Although plastics might be considered a material of the twentieth and twenty-first centuries, they are made from fossil fuels, and therefore are drawn from the deep geological time of the earth. Plastics also might be conceived of as the embodiment of artificiality, chemical formulas derived from human ingenuity, spawning materials unknown in nature. Yet they were developed in relationship to nature, as imitations of natural materials. Any number of the origin stories for plastics describe a process by which a biobased material—ivory, resin, or shellac—could not be extracted, harvested, and hunted to the degree necessary to support the emergence of capitalist consumer cultures. Therefore, the world of plastics was born as scientists sought to engineer forms that could replace the rarer, more expensive materials. But the capacity of plastics to be absorbed into the natural world—like other biobased stuff—has proved catastrophic, and many of the chapters in this volume engage with the impact of plastics on the environment and earth today.

Yet while we might conceive of plastics as solely synthetic polymers that degrade in ways that are harmful, it is necessary to understand their history in relationship to the biobased substances that they were often meant to imitate. Natural rubber—one of the iconic materials that plastics have sought to imitate—derives from trees [FIG. 1]. It was sacred to the Maya and also mixed in varying proportions with other plant matter to create various forms of rubber—from armor and sandals to rubber balls. As one historian notes, the first "polymer scientists" were the Maya. From his second voyage to the Americas, Christopher Columbus brought a rubber ball to Europe, and many of the early chroniclers of these first encounters described the qualities of rubber. Peter Martyr, who wrote a popular account of the Americas based on Columbus's travels known as *De Orbe Novo* (1555), wondered at how "with only a touch they [rubber balls] reach the stars with an incredible jump." Later in the eighteenth

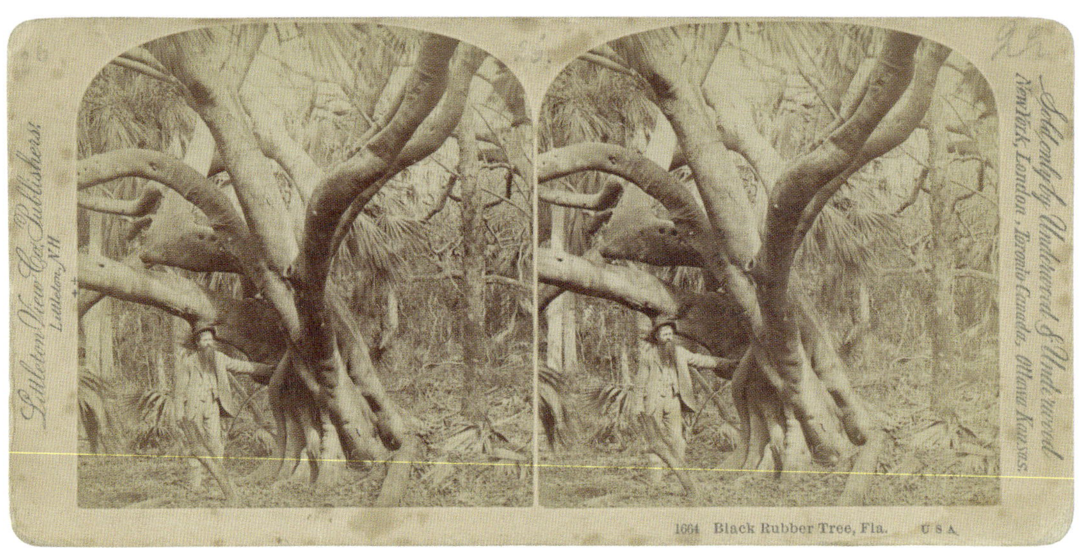

[FIG. 1]
Anonymous, *Rubber Tree in Florida*, 1891–1909. Stereography, albumen print on cardboard, 3.5 × 7 in. (88 × 178 mm). Rijksmuseum, Amsterdam.

century, Europeans encountered Amazonian people who employed rubber to make water-resistant clothing—such as rubber boots and hats. Seedlings of rubber trees from the Brazilian Amazon were carried into Asia, transmigrating an economy in rubber from Brazil to Asia and creating rubber tree plantations around the world from Brazil to Calcutta and what is today the Democratic Republic of Congo, where rubber and latex were harvested through enslaved labor. Despite the plantations of rubber trees, though, the production was not sufficient to meet the demands of communication technology when the first transatlantic cable was built in the early twentieth century, increasing the need for synthetic rubbers, which would lead to the revolution in elastomers.

Notwithstanding the production of synthetic rubber, plantations of rubber trees still exist today. Artist Eddie Aparicio traveled to a rubber tree plantation in Guatemala to make a cast of a rubber tree, noting that the air was infused with the ammonia while the toxic fumes were inhaled by Indigenous and migrant workers [FIG. 2]. Latex, the milky juice from a rubber tree, is collected in cups by slashing the tree diagonally. To prevent coagulation, ammonia is added to the containers to keep the latex fluid "running" before further processing. Aparicio's rubber tree casts formally recall other works that will be discussed in this introduction—such as Eva Hesse's innovations in the world of polymers. In turn, they also capture the tension between the natural world and the synthetic, the ingestion of chemical fumes and toxins through breathing while revealing the beautiful patterning of the tree bark on the surface of the natural latex. The natural and synthetic rubber industries exist in relationship to one another. The use value of natural rubbers stimulated the desire for inventing synthetic rubbers.

Beginning this short art history of plastics through the lens of rubber is only one origin story. Conservator Anne Gunnison begins her chapter in this volume with ivory and billiard balls at the turn of the twentieth century. There are also other places to start, such as shellac—originally used to make vinyl records. But starting with the cast imprint of the rubber tree demonstrates that, while plastics might be a relatively new invention, the creation and manipulation of natural materials to forge malleable and resistant surfaces is an age-old story. Moreover, Aparicio's cast of the rubber tree demonstrates one role of plastics in our contemporary world, which is not only as a material that furthers environmental destruction but also as a material that is central to conservation, the archive, and carrying the imprints of the natural world—holding a record of a world that is rapidly changing under the conditions of climate change. Plastics, perhaps more than any other material, are directly about the relationship between art and nature, and the desire of humans to imitate, mimic, reflect, imprint, cast, understand, and exceed the limits of the natural world as well as their place within it. There is no material more closely intertwined with the exploitation and extraction of natural resources and colonial histories of ecological destruction since 1492. Although the science of plastics began in the nineteenth century, its origin can be traced to the introduction of the rubber ball to Europe. This singular object embodies the stoking of a desire for materials that challenged the limits of the properties of nature—an ability to bounce to the stars—and the violent extraction of resources and knowledge necessary to achieve the production of commodities on a scale that would shift the ecological, economic, and cultural worlds of the modern eras.

As the conservators of the National Museum of the American Indian (NMAI) show in this volume, plastics derived from crude oil are intrinsic to the history of the Dakota Access Pipeline, a transportation network for oil that also infringes on the sacred grounds and threatens the water supply of the Standing Rock Sioux and Cheyenne River Sioux. This extraction of raw resources to support numerous industries (including plastics) is just one example of many in which the mining of resources impacts and imperils communities, and ignores the sovereign rights of Native Americans to control their land and resources—a history that started in the year 1492. Plastics, however, have also become integral to Native American and Indigenous arts, and conservators at the NMAI work to conserve plastics, and preserve the artistic legacy and knowledge of the many Indigenous cultures within the United States that now depend on plastics. Therefore, looking at plastics in relationship to art and conservation engages with one of their primary paradoxes for our contemporary world: they both play a role in our current environmental crises, and are a vital material for documenting, archiving, and preserving. They both threaten the environment and are vital to conservation.

Examining the history of plastic through the lens of art history and conservation, which is the goal of the ART/WORK series, poses a fundamental question: *What do we learn about plastics when seen through the lens of art?* How does creative practice and its intersection with the material qualities of plastics inform how we conceive of plastics as toxicity, waste, and disposable refuse while also supporting nearly every aspect of the modern world's infrastructure? This is a contradiction that nearly all the authors in this volume explore. In considering the material of plastics through the lens of art history and conservation, this volume addresses the particular challenge of plastics within the world of art, for plastics and the modern museum evolved together.

Many plastic artworks were constructed in relationship to the air-conditioned, temperate environment of the museum, which controls humidity, temperature, and light—conditions that allow plastics to remain stable within the walls of the museum. As Gunnison outlines in her

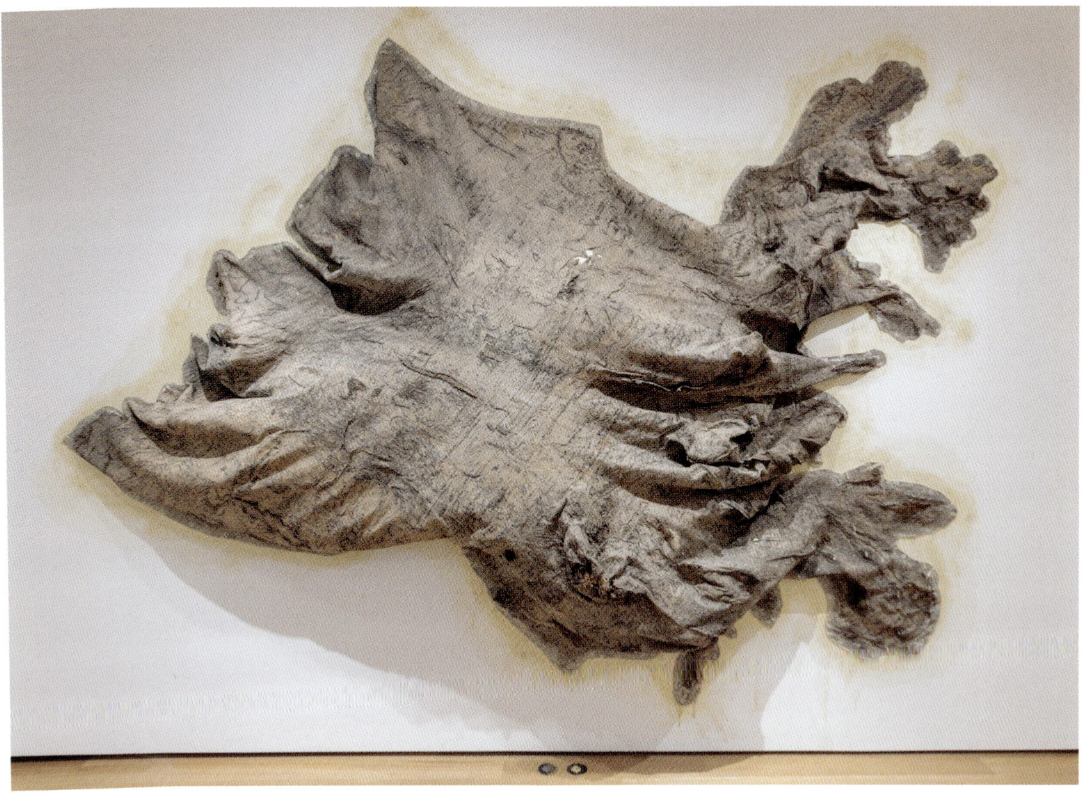

[FIG. 2]
Eddie Rodolfo Aparicio, *Pulmón #2*, 2023. Cast rubber with ficus tree residue and air from Los Angeles, San Salvador, and Massachusetts. Clark Art Institute, Williamstown, MA.

chapter, conservators and museums are dependent on plastics, from the use of adhesives in conservation to the role of plastics in creating cases, displays, and exhibition designs. Plastics were also central in building museum collections in the nineteenth and twentieth centuries, transporting artworks from distant places to European and North American collections. Gunnison discusses the use of adhesives in removing frescoes and wall paintings from distant locations to hang in European and American museums, such as the acquisition of the Dunhuang frescoes by the Harvard Art Museums in Cambridge, Massachusetts. The history of pillaging, removing, and displacing artworks across vast distances was made possible with the invention of plastics. Yet plastics also play a central role in conserving works of art for future generations.

Museums are built as repositories of knowledge and culture that will conserve objects across generations—and even perhaps millennia. But often museums themselves are ecologically unsustainable, requiring vast resources to maintain control over temperature and humidity as air-conditioned containers that depend extensively on the depletion of natural resources to preserve objects. Increasingly, many of the objects that these ecologically fraught institutions are preserving are plastics. There are many museums and initiatives working toward greater sustainability in the field of museums, and hopefully the ecology of museums will shift over the coming decades. Nevertheless, the conservation of objects and their preservation for the future often comes with a significant carbon cost, and the conservation of plastic objects is no exception. Many environmental and ecological studies of plastics focus on their inability to biodegrade while also releasing microplastics into waterways and systems, from the human body to the oceans. Art conservators, though, as demonstrated by the chapters of Gunnison and Elena Torok in this volume, work with plastics that are capable of rapid disintegration and, in some cases, combustion. While environmental scientists draw public attention to the life cycle of plastics, and their varying capacities for degrading and recycling to mitigate harm to the environment, museum conservators focus on preserving plastics for the future, particularly plastics that are unstable within museum collections.

Some of the most unstable plastic works within collections are the early twentieth-century sculptures by Naum Gabo and Antoine Pevsner, who found in plastic the ideal material to create sculpture devoid of line, volume, mass, and static form (and often color). Gabo and Pevsner sought to build sculptures that embodied a "kinetic rhythm" to capture the "perception of time." Paradoxically, however, the earliest plastics that they incorporated into their sculptures began to rapidly deteriorate, frequently yellowing and quickly becoming brittle and hard, losing the bendable, light-filled dynamism of their initial production. Instead of producing

[FIG. 3]
Antoine Pevsner, *Portrait of Marcel Duchamp*, 1926. Cellulose nitrate on copper with iron, 25.75 × 37 in. (65.4 × 94 cm). Yale University Art Gallery, New Haven, CT. Gift of Collection Société Anonyme.

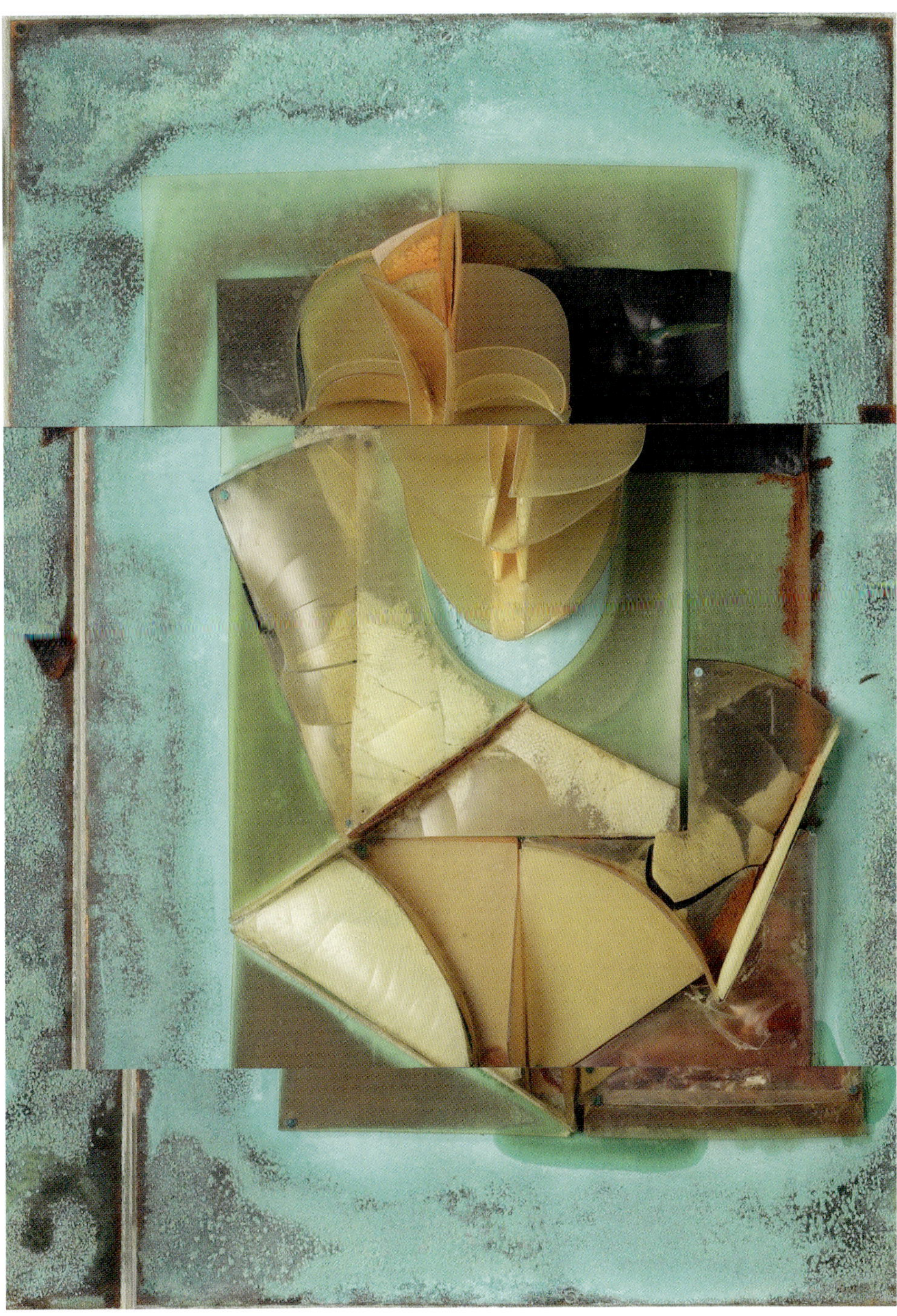

transparent sculptures devoid of line, forging constant perceptions of movement, the sculptures soon became defined by their yellowed surfaces and fragility. For example, the *Portrait of Marcel Duchamp* by Pevsner in the Yale University Art Gallery was mounted on copper after its initial mount was broken. As Torok discusses in her chapter, this new material caused a reaction with the cellulose nitrate of the plastic, triggering a rapid deterioration [FIG. 3]. The copper base quickly tarnished to a turquoise sea while the plastic became varying shades of amber. A viewer might imagine its original form: an overlapping geometry of translucent planes to create a portrait that captures the particular quality of faces to ceaselessly move and change in minute muscular shifts. The portrait has nonetheless deteriorated to a state of semifrozen brittle degradation that conveys the impact of time on materials. Pevsner and his frequent collaborator Gabo wanted to build sculptures in space that were studies in geometry, precision, and three-dimensional form, renouncing color and line as "an accidental trace of man on things." Yet Pevsner's portrait of Duchamp unintentionally embodies the "accidental trace of man on things" and inescapable ravages of time on matter. In some cases, as in Pevsner's *Torso* in the MoMA collection, the work survives in fragments and has been declared "dead." It will no longer be exhibited, although it remains in the collection for research and archival purposes. In her chapter in this volume, Torok explores the impact that the fragility of their works had on Pevsner and Gabo along with their legacy.

Other artists, however, conceived of the degradation of plastics and incorporated this into their fascination with the material. Hesse, whose first major solo exhibition was called *Chain Polymers* (1968), recognized the fragility of her works. As Hesse stated in an oft-cited interview from *Artforum* in 1970, "And then the rubber only lasts a short while. I am not sure where I stand on that. At this point I feel a little guilty when people want to buy it. I think they know but I want to write them a letter and say that it's not going to last. . . . Life doesn't last; art doesn't last." While her most iconic work from that exhibition, *Repetition Nineteen*, is now relatively stable in its current state, it has also achieved a patina of yellow infusing the fiberglass and resin [FIG. 4]. The yellowing was not intrinsic to their original appearance, but it further enhances the embodied qualities of contingency and imperfection that the installation suggests in the arrangement of the vessels scattered across the floor, bending toward and away from one another. Hesse's diaries demonstrate that she attended a lecture series in New York City for artists, Experiments in Art and Technology, in which scientists from Bell Telephone Laboratories gave talks on topics such as "Polymers" and "Casting, Molding, and Control." Her surviving notes on the lectures reveal that she was invested in the chemistry of polymer science. Whereas Pevsner did not intend for

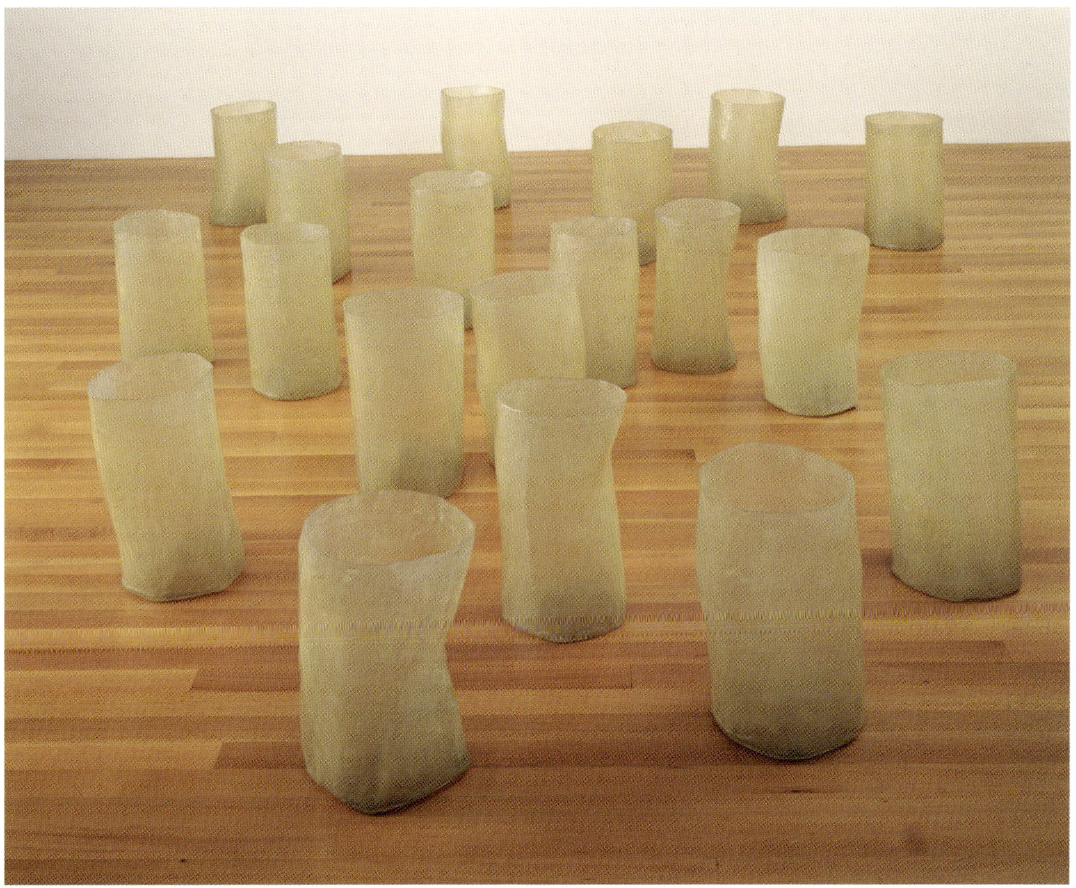

[FIG. 4]
Eva Hesse, *Repetition Nineteen III*, 1968. Fiberglass and polyester resin, nineteen units, each 19 to 20.25 in. (48 to 51 cm) and 11 to 12.75 in. (27.8 × 32.2 cm) in diameter. Museum of Modern Art, New York.

his sculpture to morph and become brittle, Hesse embraced the shifting qualities of plastic materials. Nevertheless, both artists belong to a certain genealogy of artists whose careers and work were shaped by the particular qualities of plastic. For Pevsner, the material held possibilities for creating form devoid of line, color, and volume, while for Hesse her works consider the limits and beauty of bodies in the wake of the world wars, Holocaust, and postwar industrialization. In *Repetition Nineteen*, the vessels bend, turn, and lose their geometric perfection and clear transparency in favor of a slight deformation that allows for a reflection on as well as rejection of smooth impossible surfaces.

Still, although certain artists explicitly engaged with plastics as a material along with its limits, to write a short history of plastics in art is almost an impossible task, as David Joselit demonstrates in his chapter. Beyond artists such as Pevsner and Hesse who studied the material properties of plastics, polymers touch on almost every aspect of modern and contemporary artistic production. Acrylic paints are plastic. Film is fabricated from plastic. Most artists' materials today would be impossible without the plastics industry. While single-use plastics and the petrochemical industry are an emblem of the destructive force of human innovation and design, plastics are the surface and material that defined twentieth- and twenty-first-century artistic innovation too. Painting, sculpture, film, and photography—they are all mediums defined by plastics in the twentieth and twenty-first centuries. This is the complication with thinking about the specific materiality of plastics—their *plasticity* allows them to exist within a wide variety of matter—from paint to film to sheeting. This variety and variability make the specificity of plastic as a material nearly impossible to consider. Plastics touch every aspect of artistic production whether it is a painter working in acrylics, sculptor working with resin (as will be seen in this introduction), or photographer working with celluloid film. Plastics shift their specific quality across practically all media of modern and contemporary art.

Unlike bronze, marble, tempera, or even paper, plastics are too expansive a material to productively engage a specific material history. As artist Kevin Beasley shows in his chapter in this volume—which is an exploration of the plastic materials that structure the space of his studio—plastics are not only a material with which artists work but the atmosphere that we all breathe as well. In turn, in her chapter, Jennifer Gabrys discusses the atmospheric qualities of plastics and dangers they pose to inhalation. When the world is viewed through the lens of microplastics, plastics run through our systems and bodies, restructuring our biological forms. While artists have no doubt inhaled marble dust and paint fumes for centuries, there has never been such a full-scale alchemical transformation of the natural world through the micro- and nanoworld of plastics. In this way, plastics are difficult to write about because they are atmospheric.

As any conservator of plastics, film, or archives know, there is a specific pungent smell of camphor or vinegar that hits one's nostrils and lungs when one encounters a plastic work that has begun to degrade. The smell when viewing a Pevsner sculpture within a conservation lab comes with an olfactory punch that makes evident how plastics infiltrate the air. This physical atmosphere of plastics is evoked in the work of Aparicio that started this introduction and his discussion of the ammonia in the air of the rubber tree plantations—chemicals meant to keep the natural latex flowing for further processing. There are also other contemporary projects that examine this consideration of "aspiration" that has derived from the rise of plantation economies.

Beasley's exhibition *A View of a Landscape*, which ran from December 15, 2018, through March 10, 2019, at the Whitney Museum, directly engaged with the atmosphere of environmental degradation and its impact on bodies. In this work, Beasley restored the motor of a cotton gin and installed it into a soundproof chamber [FIG. 5]. In the space of the gallery, the motor whirred within the chamber, although the gallery itself remained silent. In the adjacent room, Beasley remixed the sounds of the cotton gin motor to create music and ambience—space for living within and with the legacy of the plantation economy while transforming it into a new ontological possibility, transmuting the sounds of the cotton gin into music, ambience, and space that could interpenetrate into bodies within the aesthetic space of the museum. While Beasley's installation might seemingly only touch on the history of cotton in the United States, cotton is central to the history of plastics and film too. In the nineteenth century, it was discovered that treating cotton cellulose with an acid mixture (various acids and combinations of alcohol, ether, and camphor, among other "plasticizers") created a liquid that formed into a transparent sheet ultimately known as celluloid. Yet the earliest forms of celluloid were extremely dangerous, often derived from "gun cotton," or a military-grade gunpowder. Gunnison discusses this history further in her chapter, but one of the most important and influential archival surfaces of the twentieth century—cellulose nitrate film—is flammable and combustible, capable of burning without oxygen. It is also a surface intrinsic to the history of the cotton plantation economy.

Yet in the work of Beasley, plastics often appear to archive for the future. Beasley sometimes uses polyurethane resin to "fossilize" cotton surfaces and materials, endowing a veneer of survival beyond the life-span of the garments. Whereas textiles are famously sensitive to conditions of light and temperature (particularly in humid climates such as the US South), Beasley employs plastics to preserve these garments and

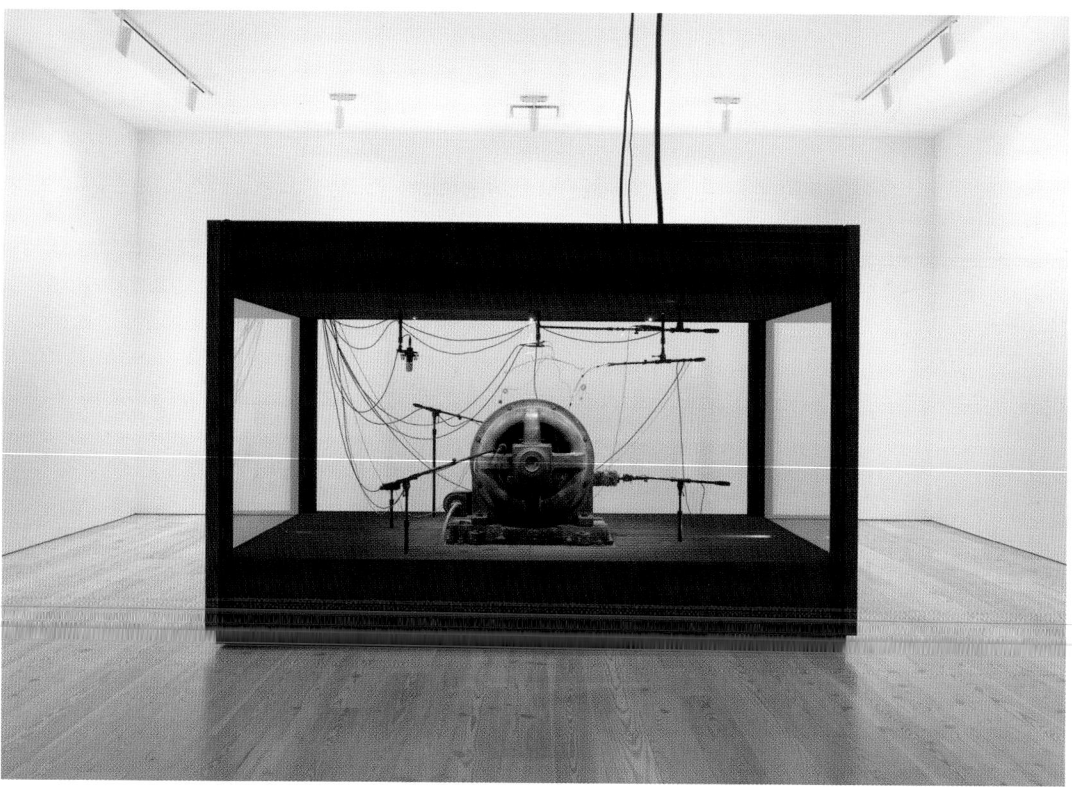

[FIG. 5]
Kevin Beasley, *A view of a landscape: A cotton gin motor* (detail), 2012–18. Chamber: GE induction motor, custom soundproof glass chamber, anechoic foam, steel wire, monofilament, cardioid condenser, microphones, contact microphones, microphone stands, microphone cables, AD/DA interface. Listening Room: Custom speaker system, subwoofers, amplifiers, AD/DA interface, ethernet switch, mixer, modular synthesizer, equipment racks, wood table. Vitrine: 96 × 120 × 96 in. (243.8 × 304.8 × 243.8 cm).

other materials that might otherwise degrade and disappear [FIG. 6]. In preserving cotton, Beasley demonstrates that conservation is also central to the history of plastics, which meets the instinctual human impulse to archive for the future, so that the impression that we have left on this world does not disappear. Nevertheless, this desire has pushed the world to an unsustainable extreme so that the world itself, or at the very least our place within it, has appeared increasingly acutely finite.

Today, the history of plastics is inseparable from the history of this earth. In his discussion of plastics in his artist's studio, Beasley makes evident the integral role of plastics in both daily life and artistic practice while acknowledging how they infect the air that we breathe. Artists and conservators take a particular vantage point in trying to understand this material—frequently aware of the environmental problems posed by plastic but also invested in its design, its possibilities, and the ways in which it reflects like no other material the complexity of history after 1492. From the perspective of art history, plasticity as a concept defines modern and contemporary artistic production, although as Zakiyyah Iman Jackson shows in *Becoming Human: Matter and Meaning in an Antiblack World*, plasticity is also an "anti-Black mode" that puts a demand on Blackness and Black bodies to be "infinitely mutable," particularly in regard to sex, gender, and reproduction. Moreover, the plasticity assigned to the human body is a construct that Bess Williamson explores in her chapter, considering how the world of plastic prosthetics can force bodies into idealized images of what it means to walk, sit, eat, and exist—again, considering the conceptual world of plasticity as a violence that is often enforced from without. Still, as Williamson argues, plastic's malleability, disposability, and accessibility sustain modes of living for people with disabilities, or certain chronic conditions. Like many of the chapters in this volume, Williamson considers both the ontological violence of plasticity as a concept and the material of plastics overall, demanding recognition for the necessity of the world that plastics make possible. As Williamson points out, banning certain plastic materials—such as straws—distracts from the larger structural problems that threaten both the environment and communities most at risk of environmental catastrophe.

Williamson's chapter touches on one of the most important histories of plastics too—namely, their centrality to twentieth-century design and modernism. As Jason Weems contends, Charles and Ray Eames were innovative during World War II in creating biomorphic furniture with swelling curvatures to mold to the human body and making leg splints for soldiers. This fascination with molding form to the human body would ultimately impact their furniture design and such iconic forms as the Eames chair, a piece of furniture that contours itself to the human body. But as art historian Kristina Wilson asserts, midcentury design and

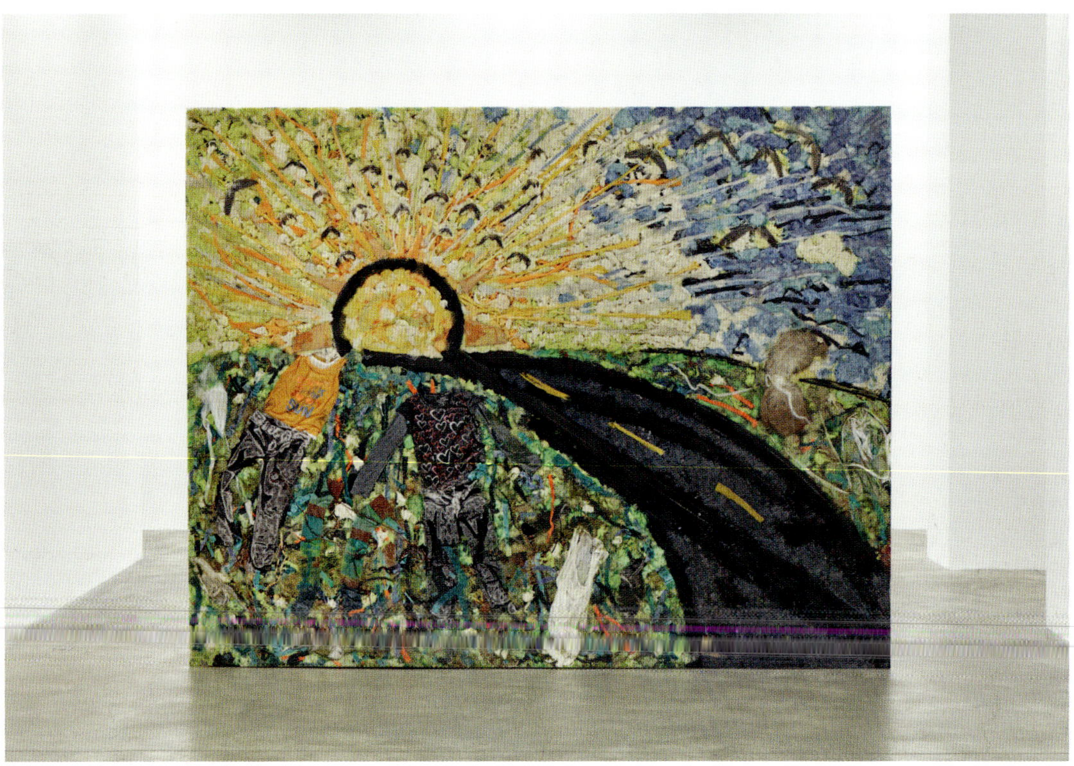

[FIG. 6]
Kevin Beasley, *The Road*, 2019. Polyurethane resin, raw Virginia cotton, Virginia soil, Virginia twigs, Virginia pine needles, housedresses, kaftans, T-shirts, du-rags, altered housedresses, altered kaftans, altered T-shirts, altered garments, altered tires, scarf, guinea fowl feathers, down feathers, copper, jewelry, shoelaces, mobile phone, burlap satchel, windshield wipers, altered African fabrics, socks, Timberland boots, aluminium, and steel, 96 × 120 × 10 in. (243.8 × 304.8 × 25.4 cm). Whitney Museum of American Art, New York.

objects (made possible through plastics) frequently construed their forms in relationship to an "ideal" body. Midcentury design too often assumes the contours of a thin, able (and usually racialized-as-white) body that can slip into the cocoon of its lines and hollow spaces. The bodies that do not fit within that model are left to shift, spill over, hunch, slump, and exist in perpetual discomfort.

ART/WORK as a series is meant to make histories of conservation legible to nonspecialists while also engaging in the various ways in which writing about art history and methods is informed by knowledge of the properties of a specific material. Plastic, in particular, poses important challenges and promises to this premise. Gunnison shows that despite the typically held view that plastics have long shelf lives and cannot decompose, there are unstable categories of plastics. Within the case studies, Torok introduces the handling of a canonical piece of twentieth-century plastic sculpture from the view of a plastic conservator while the conservators from the NMAI discuss handling the presence of plastics within their collections. Exploring contradictory aspects of plastics, Williamson explains the centrality of plastic as a material to design history and disability studies. Finally, Gabrys illustrates how plastics exist in the very air that we breathe, while Beasley, an artist dedicated to questions of aspiration, breath, and materials (such as cotton and plastic), walks the reader through his studio.

Like all the volumes in this series, this is not a comprehensive history of plastics. It is an introduction to thinking about plastics in art history. Returning to the fascination with the rubber ball, plastics and plasticity are harmful when they presume that the earth, our resources, our bodies, and all species are infinitely malleable, reproducible, and capable of defying the limits of gravity and space. The production of plastics supported a vision of a world in which natural resources could be created if they became scarce, and we could build materials that would not age, tear, or be destroyed. It is not so much plastics themselves that have been destructive but rather the fantasy that they have generated. Yet as art conservators recognize, plastics are limited, and they are also central to the act of conservation itself in the twenty-first century. The material qualities of plastics explored in this volume are by thinkers who consider what it means to fossilize and preserve our plastic world for an uncertain future, recognizing that whatever happens, it is clear that our present will be seen in the future through plastic and its refuse.

Conservation and Plastics

Anne Gunnison

In 1878, the Celluloid Piano Key Company published the musical score *Celluloid Grand March Op. 4* for the pianoforte, composed by Charles Wels, who is listed as the company's vice president [FIG. 7]. The score is not so much concerned with compositional innovation but rather material invention, as the last few pages are an advertisement for a new material, celluloid ivory, marketed as a replacement for elephant ivory for piano keys, organs, and melodeons. There are testimonials from organ and piano company proprietors who extol the virtues of the celluloid keys. In bold, the ad claims, "In fact, under every condition and circumstance, for Piano, Organ, and Melodeon Keys, the Celluloid Ivory is better than Elephant Ivory, and is speedily and surely supplanting it." And so the great march of the plastics begins.

Celluloid ivory is cellulose nitrate, a semisynthetic plastic, developed and patented in England by Alexander Parkes in 1865. His attempt to take this market under the name of Parkesine failed by 1868. Parkes's US counterpart, John Wesley Hyatt, was simultaneously seeking to discover an alternative for elephant ivory, which was used to make billiard balls. Ivory was scarce and expensive, and Hyatt's endeavor was made more appealing by the promise of a $10,000 reward by the billiards company Phelan & Collender to anyone who developed a replacement for ivory. Hyatt was successful in both developing cellulose nitrate as a material and making it commercially viable. By 1868, Hyatt's Albany Billiard Company was in business. By 1870, Hyatt and his brother had established the Celluloid Manufacturing Company in Newark, New Jersey. Beyond billiard balls, celluloid, the name Hyatt trademarked in 1871, was used primarily as an imitative material for more expensive and rare materials (ivory, tortoiseshell, and horn), to manufacture shirt collars and cuffs, hair combs, and home goods.

By 1968, one hundred years after the founding of the Albany Billiard Company, plastics had become something completely different and yet wholly the same. The formulas and chemical compositions were different and greatly expanded, but the principle of plastic as a malleable replacement for something natural was familiar. *Motor Trend*'s car of the year in 1968 was the General Motors Pontiac GTO, the first year it was outfitted with the Endura bumper. Instead of a metal bumper, which had graced the front of cars since auto manufacturer Henry Ford's early models, the Endura bumper was made of high-density urethane elastomer—that is, a plastic. Ford himself had developed a plastic car in 1940, though the bumpers were still in fact metal. (And only one of Ford's plastic cars was ever made.) In a television advertisement for the GTO—as perhaps a nod to Ford's own demonstration of hitting his plastic car with an ax, causing no damage—an actor in a white tuxedo approaches the GTO with a crowbar and repeatedly hits the front bumper, causing no dents or cracks. After throwing the crowbar to the ground, the actor declares, "This is a brand new 1968 Pontiac GTO. It has a new bumper. It's the most revolutionary new bumper, since . . . bumpers. Seems like everything those Pontiac engineers touch turns to great." In another version, a team of engineers, white men in white lab coats, takes turns trying to smash the bumper with a crowbar while the engineers' colleagues frantically and aggressively yell encouragement. They are unsuccessful. It does not smash, crack, or break. Nevertheless, although the bumper may survive a fender bender, it could not survive the ravages of the elements. The bumper was known to crack and fade, especially in hotter climates, and Pontiac had to release a repair kit.

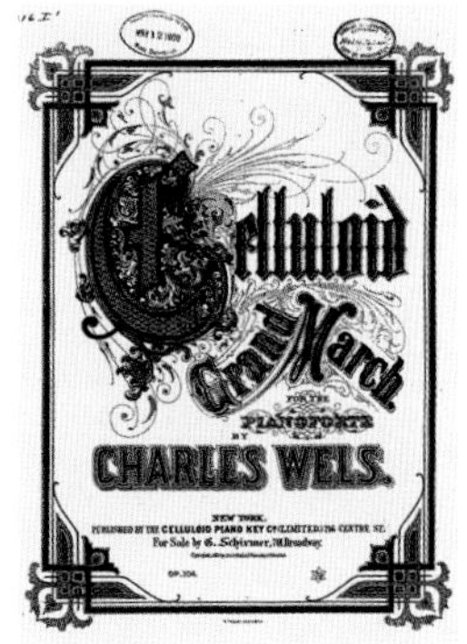

[FIG. 7]
Charles Wels (composer), *Celluloid Grand March*, New York, 1878. Celluloid piano key. Library of Congress, Washington, DC.

That same year, 1968, the movie *The Graduate* was released. In a now-iconic early scene, recent college graduate Benjamin Braddock, played by Dustin Hoffman, is pressed about his future career and cornered by a Mr. McGuire, who has some advice:

> *Mr. McGuire*: I want to say one word to you. Just one word.
> *Benjamin*: Yes, sir.
> *Mr. McGuire*: Are you listening?
> *Benjamin*: Yes sir, I am.
> *Mr. McGuire*: Plastics.
> *Benjamin*: Exactly how do you mean?
> *Mr. McGuire*: There's a great future in plastics. Think about it. Will you think about it?

For Braddock, plastics are banal and representative of a pretty bleak future for him and his generation. Plastic is both a material and a concept. In the case of *The Graduate*, the prospect of a career in plastics suggests the ennui of mainstream middle-class America that the contemporaneous counterculture in the United States sought to reject. This was also the year that the Plastic People of the Universe (PPU), a Czech band, formed in Prague, two months after Soviet troops invaded Czechoslovakia on August 20, 1968. The arrest of PPU band members in 1976 partially influenced, at least in popular lore, the drafting of Charter 77, a petition written by Czechoslovakian writers and intellectuals, including future Czech Republic president Václav Havel, demanding the Communist Czechoslovakian government recognize basic human rights. PPU's name was inspired by the 1967 song "Plastic People" by Frank Zappa and the Mothers of Invention. Zappa's lyrics make it clear that plasticity is not something to aspire:

Take a day and walk around
Watch the Nazis run your town

You think we're singing 'bout someone else
But you're plastic people
Oh baby, now you're such a drag

Me see a neon moon above
I searched for years, I found no love
I'm sure that love will never be
A product of plasticity

In Zappa's lyrics, plasticity is indicative of something much more sinister than Braddock's depressive future. It is equated with fascism and those who are unquestioningly complicit with it, but the PPU took the concept of plastic people to represent the antiestablishment.

This duality is a defining characteristic of plastics: it can be both revolutionary and conformist. Plastic is a semantic monolith, representing a material that has been developed to be a physical and ideological chameleon. Plastics are endless permutations of distinct chemical reactions and formulas, all categorized as plastic. Plastics are duplicitous, as their very origin story is rooted in pretending to be something that they are not. Is plastic long-lasting, or does it degrade easily? Both. Is it organic or inorganic? Primarily derived from organic materials (petroleum and cellulose), plastics in general do not biodegrade or decompose like natural organic material but rather disintegrate into smaller and smaller pieces, which remain in themselves plastic. Plastics can be rigid or flexible, transparent or opaque. They can be a problem. They can be a solution. The over 150-year development of plastics from an alternative to ivory and tortoiseshell to a universally pernicious substance can be traced through patents, advertisements, scientific papers, the history of industrialism and colonialism, environmental reports, design trends, and popular culture in a way that is seemingly incomparable in scope with any other material. And this is reflected in objects of cultural heritage, which have been acquired by museums as historical objects, artwork, design objects, scientific specimens, and more.

Though plasticity and plastic are concepts, plastics are also very much a specific material, from hair combs to reels of film to monumental sculpture. The materiality of plastic in an object cannot be ignored when understanding the concept of the object. Plastics can be some of the most stable objects in a museum collection but equally the most fragile and prone to severe degradation. Conservators' decisions and hands-on technical skills aim to preserve as well as care for tangible history and art objects. To make decisions about the appropriate methods of preservation of this cultural property, a conservator must have a basic understanding of the materials. Plastics, however, come into the museum as so many formulas, shapes, properties, and manufacturing histories that identification and understanding can be difficult. They can be commercial products with proprietary recipes. Some plastics were marketed under specific names, which can both reveal and obfuscate their type as well as function. It is no wonder that this material is aggregated into a common term: plastic.

The Fourth Kingdom

To be a conservator, tasked with studying and preserving cultural heritage objects, is to be constantly playing a game of twenty questions: What is this object? Where did it come from? Who made it? What is it made of? How was it made? And to be a conservator in a museum that has collected objects from the past 150 years is to face the conundrum of plastic. Plastic? Animal, vegetable, or mineral kingdom? Or is it its own kingdom all together? The so-called fourth kingdom is a notion introduced in a marketing movie filmed in 1937 by the Bakelite Corporation, a plastics company: "Our modern complex industrial world has turned elsewhere to fulfill its needs. Turned to a fourth kingdom. A kingdom of scientific research, a new domain of man's own creation. A world of primary substances which never existed before."

The materials in this fourth kingdom are hard to comprehend because plastic, a name that can encompass so much, is a product of a complex industrial process, and it can be difficult to determine just what it is. It is often easier to simply describe it as plastic.

What Are Plastics?

They start small. Monomers are molecules that are bonded to other identical molecules. During the polymerization process, monomers react together to form polymers. Polymers are chemical compounds that are long repeating chains of molecules. Plastics are polymers, though not all polymers are plastic. While this discussion is primarily focusing on plastics as manufactured or synthetic substances, there are natural polymers. These include bitumen (a tar that is the thickest form of petroleum), shellac (secretions of a lac bug), gutta percha (a latex from the *Palaquium gutta* tree), rubber (from *Hevea brasiliensis* trees or *Landolphia* vines), and cellulose (plant matter) as well as horn and tortoiseshell (keratin).

Semisynthetic polymers are made when a natural polymer like cellulose is chemically reacted with an organic compound. An example is cellulose nitrate, formed when cellulose fibers are treated with a solution of nitric acid. Hydroxyl groups on the cellulose are replaced with nitrate groups. In the case of John Wesley Hyatt's celluloid, the addition of a solvent and camphor, a white crystalline solid extracted from a camphor tree, to the nitrated cellulose prevented significant shrinkage while drying.

Synthetic polymers are those that are derived from petroleum, natural gas, or coal, though there are biobased polymers, for which a percentage of the monomers are from natural materials like corn or starch. The first truly synthetic plastic, phenolic resin, converted from phenol (a coal product) formaldehyde, was discovered in 1907 by Leo Hendrik Baekeland. He lent his name to this plastic, which was called Bakelite. Just as cellulose nitrate was used as a replacement for other natural polymers and materials, phenolic resin was developed as a replacement for shellac (often used for music records). But even in Bakelite's infancy, Baekeland recognized it as a material that could have many applications. Synthetic plastics are further broadly categorized as thermoplastic or thermoset. **Thermoplastics**, because of their polymer structure, are able to be remelted and reformed. **Thermosets**, because their polymer chains cross-link on setting, can only be heated and formed once. Museum objects can represent natural, semisynthetic, and synthetic plastics. Most plastic objects made after the 1920s will be synthetic plastics, however, representing an incredible range of plastic families: polyolefin, vinyl, acrylic, styrenic, polycarbonate, phenolic, and epoxy, to name a few. And most of these will have been derived from petroleum processing.

Natural gas and crude oil are mixtures of hydrocarbons. These hydrocarbons are organic compounds of hydrogen and carbon molecules derived from decomposed plant as well as animal life from hundreds of millions of years ago that have undergone heat and compression through geological time. After sources of natural gas or crude oil are drilled and pumped, the raw material is refined, and the mixture of hydrocarbons that comprise the gas or oil are separated into groups. Two of these types of hydrocarbons are **naphtha**, which contains ethane and propene (also called propylene) and is derived from crude oil, and **ethane**, primarily derived from natural gas. Ethane and propene are heated to a high temperature in a processing plant called a cracker. This name comes from the function of the plant: to crack the molecules of the hydrocarbons under high temperature, turning them into the monomers ethylene or propylene. These monomers are then polymerized with different types of chemical catalysts to engineer synthetic polymers with varying properties including polyethylene, polyester, polypropylene, polyamide, polystyrene, and polyvinyl chloride, among many others. The long chain polymers serve as the backbone of the product and additional characteristics are introduced to the polymer through additives. These can include antiaging through heat stabilizers, ultraviolet (UV) absorbers, and antioxidants; antistatic, blowing, and cross-linking agents; colorants; fillers; fiber; flame retardants; lubricants; and plasticizers. Plastics therefore represent endless permutations and recipes.

After polymerization, plastics can take the form of small pellets (< five millimeters) called preproduction plastic pellets, or colloquially, nurdles. These pellets are then used to manufacture countless plastic products through extrusion, melting, and molding. While these polymers are "organic" in that they have a carbon backbone, they are unable to be broken down readily in the environment because they are long chain polymers containing thousands of monomers, hence their strength. Currently, the most dominant and inexpensive types of polymers that are produced globally are polyethylene and polypropylene, because they are easily sourced from natural gas, resistant to degradation by water, air, or solvents, and easy to shape into a variety of objects, though mainly they are used for packaging. Given that the history of the development of synthetic plastics is well-documented, plastics can be used as a dating tool for objects. If the date of manufacture is known, the type of plastics may be identified based on the plastics available at the time or by its physical characteristics. Or sometimes it is as easy as looking at the symbols and/or brand names on a plastic object to identify the type.

In 1988, the Society of the Plastics Industry Inc., founded in the United States in 1937 in response to a growing number of recycling programs, developed a consistent code called the Resin Identification Code that identifies the plastic resins contained in products. In 2008, the administration of this coding system was turned over to the American Society for Testing and Materials International, an international standards company, which has developed the continually updated Standard Practice for Coding Plastic Manufactured Articles for Resin Identification. The numbers 1 through 7 on products, usually shown in a triangular recycling symbol (previously composed of three arrows, but now an equilateral triangle), identify the following: (1) polyethylene terephthalate (PET) bottles, cups, and other packaging; (2) high-density polyethylene bottles, cups, milk jugs, and so on; (3) PVC (or vinyl) pipes, siding, flooring, and so forth; (4) low-density polyethylene plastic bags, six-pack rings, and tubing; (5) polypropylene auto parts, industrial fibers, and food containers; (6) polystyrene utensils, Styrofoam, and so on; and (7) others, including materials made with more than one resin from categories 1–6 as well as acrylic, nylon, polycarbonate, and polylactic acid. The number also indicates the relative ease and cost of recycling the material, with 1 being the most readily recyclable and 7 the most difficult.

Plastics as Museum Objects

It is difficult to determine, historically, when plastic objects first entered museum collections. And it would be impossible to itemize the types of objects made from semisynthetic and synthetic polymers that enter collections: cups, computers, paintings, cigarette holders, beads, clothing, chairs, sculptures, packaging, printed posters, and toys. For a conservator, it is easiest to think of how and why plastics are acquired, and at what point in the lifetime of the material, in relation to its manufacturing and use history, it enters the museum. This life cycle roughly follows the steps below:

1. Raw material acquisition: oil, natural gas, or biomass production
2. Material processing: creation of monomers and reagents for polymerization, and subsequent polymerization and compounding to create bulk plastic materials
3. Manufacturing and assembling: molding, extruded or other conversion of bulk plastic compounds, and any secondary products
4. Use life: the employment of the plastic product by the consumer, handling the product, and the effects of use on the plastic
5. End of life: the recycling or reuse of the product/material before it is disposed
6. Disposal: the final resting place of the plastic product, such as a landfill, incineration, or compost bin

Museums would not acquire plastics in the stages of raw material acquisition—as oil or natural gas. Yet the later stages of production have been archived by certain museums. In Sydney, at the Technological Museum (now the Powerhouse Museum, a branch of the Museum of Applied Arts & Sciences), for example, Arthur de Ramon Penfold, a former industrial chemist who worked as curator and later director of the museum, began in the 1930s to acquire specimens of plastic raw materials and finished products to illustrate the development of synthetic plastics. He requested samples from manufacturers and acquired objects that would represent this burgeoning fast-growing field. In 1934, the Sydney Technical College and the museum displayed the first Plastics Industry Exhibition in Australia. Following the success of this exhibition, a permanent display of plastics was established at the museum, described by the *Sunday Telegraph* as "the best display of plastics and fibres in the world show(ing) the complete history of plastics from first experiments to the latest developments."

In some cases, plastics enter the collection as modified forms of new plastics that have never had a use life as their production intended. For instance, Canadian artist Brian Jungen [see Heald and McHugh, this volume] purchases new plastics such as suitcases, chairs, Nike sneakers, and synthetic basketball jerseys, and then manipulates these objects to create artworks. The plastic objects from which these finished sculptures originated have never been used as intended, and their use life becomes defined by their manipulation in the artist's studio.

In other cases, plastics that have gone through the complete life cycle, from raw material acquisition to disposal, are then salvaged for another use life. This is seen in the work of contemporary US artist Duke Riley [FIG. 8]. He uses plastics that have been expended and discarded. Riley removes them from the disposal process and creates works of art. In mirroring the history of plastics—as a material invented to replace valuable animal products such as horn, ivory, and tortoiseshell—Riley replicates scrimshaw, ink drawings etched into ivory, including walrus tusks and whale teeth, an artwork primarily made by whalers and seafarers in the nineteenth century. But instead of using these endangered materials, Riley replaces the tusks and teeth with discarded plastics—detergent bottles, toothbrushes, and other waste he has collected on beaches—to etch ink drawings onto their plastic surface.

In other capacities, there are museum objects that are not readily thought of as plastics, including motion picture film and synthetic painting media, semisynthetic and synthetic polymers, which must be taken into account as such. Motion pictures are primarily thought of as what is projected onto a screen. Yet this projection is only possible because of the composition of film, which was initially made from cellulose nitrate (from which derives the term *celluloid* for motion picture film) but was gradually replaced by cellulose acetate and now is primarily PET. Synthetic paints were developed to overcome some of the limitations of oil paints, organic binders, and natural resin coatings and varnishes. Starting in the 1920s, fast-drying cellulose nitrate lacquer, which could be sprayed and not brushed on, began to be used in the automobile industry to replace slow-drying oil resin paints. These paints were used by artists too, including in the 1930s by Mexican muralists like David Alfaro Siqueiros, who took advantage of the ability to spray these lacquers to paint larger surfaces [FIG. 9].

Alkyd paints, also introduced in the 1920s, were polymerized polyester with fatty acids. Though drying more slowly than cellulose-nitrate-based

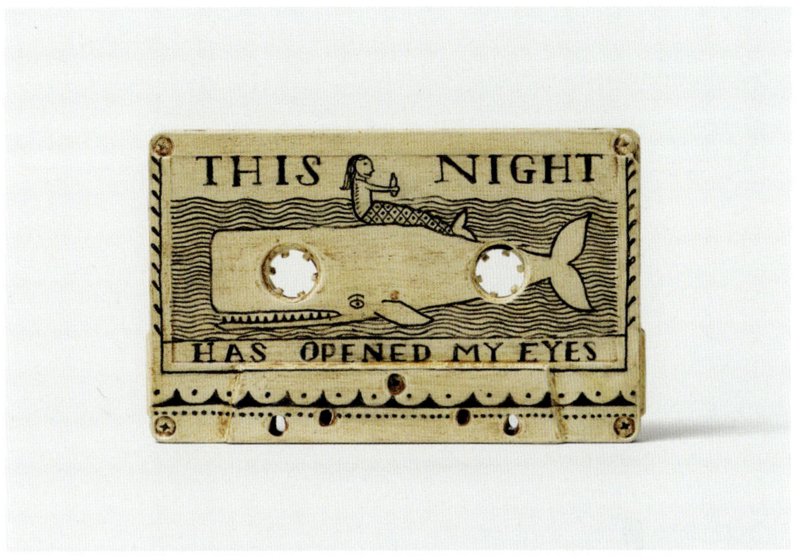

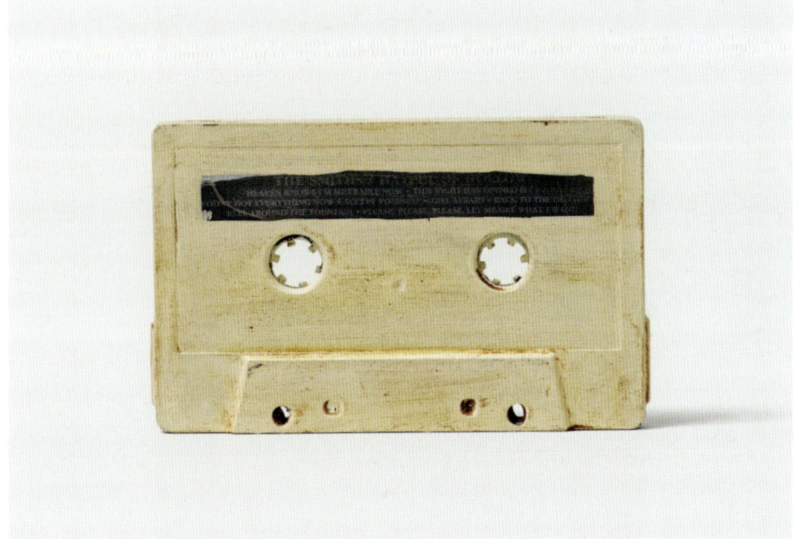

[FIG. 8]
Duke Riley, *No. 363 of the Poly S. Tyrene Memorial Maritime Museum*, 2023. Painted salvaged plastic, ink, and wax, 25 × 4 × 0.25 in. (63.5 × 10.2 × 0.6 cm). Duke Riley Studio, Brooklyn, NY.

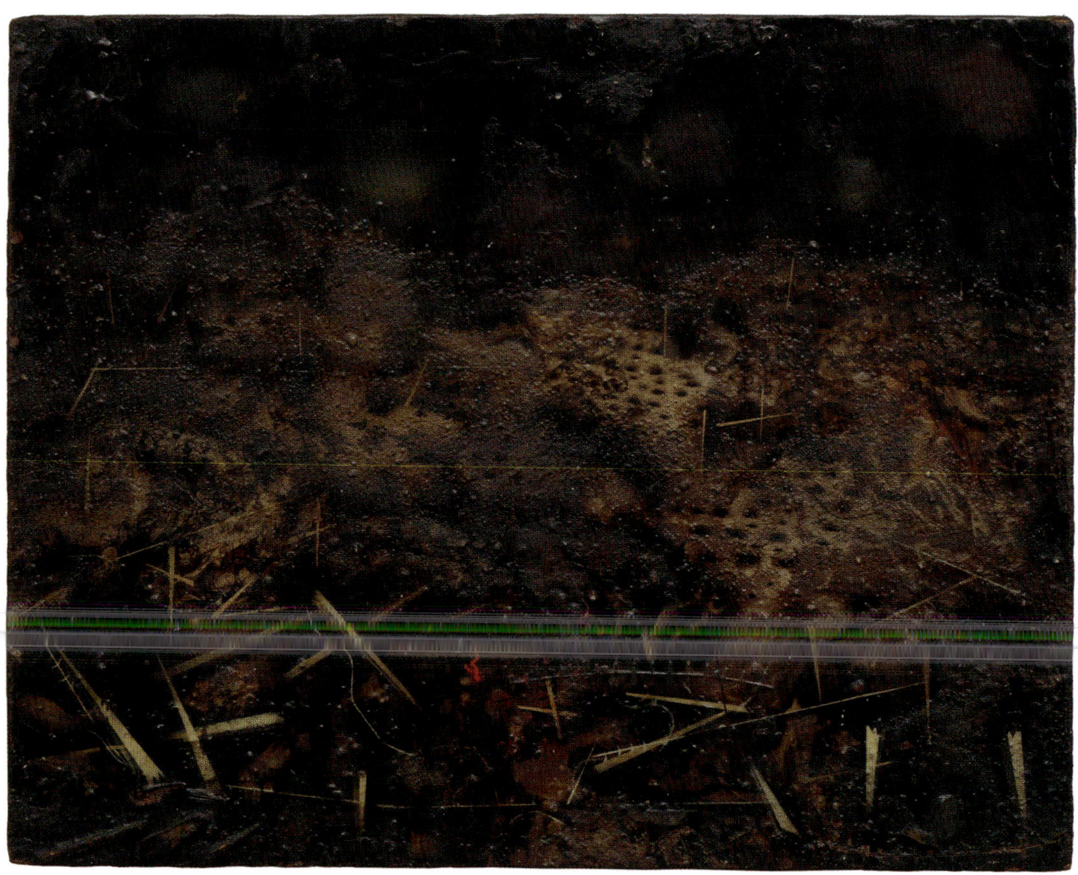

[FIG. 9]
David Alfaro Siqueiros,
Cosmos and Disaster,
1936. Lacquer, pyroxylin
sand, and wood on copper
mesh over plywood, 23.9 ×
29.9 in. (608 × 761 mm).
Tate, London.

paints, the alkyds behaved much like oil paint but dried more quickly and were thinned using low-aromatic hydrocarbons such as mineral spirits. Alkyd paints used for painting appliances, cars, and trains, and marketed as house paints, were adopted by artists in the 1940s including Jackson Pollock, who used cellulose nitrate lacquers as well. In the 1940s, mineral-spirit-based acrylic resin paints, developed by Leonard Bocour and Sam Golden, marketed as Magna, were developed specifically for artists' use. Roy Lichtenstein and Morris Louis were loyal users. In the 1990s, Magna, which was then produced by Zipatone of Chicago, was discontinued. By the 1950s, water-based acrylic polymer emulsions were, however, widely available to artists. These types of paints dry quickly and are flexible when dry as well as generally resistant to fading and yellowing. Therefore although the production of plastics might typically be associated with sculpture, three-dimensional works, or even film, their presence in museums extends into two-dimensional media and painting, as acrylic paints are commonly found as the main media for paintings executed from the mid-twentieth century onward.

Plastics as Problem

If the intention of a museum collection is to be the final resting place of the plastic product, it is the conservator's job to prolong its life, even if it is in its final stage of degradation. This can be a challenge. Despite the apparent resilience of plastics, they are susceptible to degradation by the forces to which they are exposed every day: light, heat, moisture, and even oxygen. While many plastics are relatively stable (particularly if stored in optimal conditions of light, humidity, and temperature), there are five plastics that have been identified as "problem plastics" in museum collections: cellulose acetate, cellulose nitrate, plasticized PVC, and polyurethane (PUR) foams and PUR elastomers.

The **cellulose nitrate** developed by Parkes and Hyatt was a descendant of German Swiss chemist Christian Friedrich Schönbein's 1840s' invention, guncotton or pyroxyline, cotton treated with sulfuric and nitric acid. As the name suggests, guncotton was highly combustible. It was intended for military use but was incredibly unstable and prone to exploding spontaneously. When dissolved in organic solvents, the pyroxyline becomes a lacquer-like solution. A form of this is called collodion, still used in wet plate photography. Parkes and Hyatt's cellulose nitrate is also cotton treated with sulfuric and nitric acid, or just nitric acid, but with additions of plasticizers, particularly camphor to create a more stable, moldable material. The proclivity toward instability remained, though.

The more nitrated the cellulose is—that is, the number of nitrate groups that replace the hydroxyl groups on the cellulose during the synthesizing process—the less stable the cellulose nitrate. Movie film was especially highly nitrated, having a nitrogen content of 12 to 14 percent, close to that of guncotton, while other cellulose-nitrate-based articles, like housewares, were generally 11 percent and thus more stable. The stability of cellulose nitrate depends a lot on how it was manufactured—perhaps most critical, the thoroughness of the washing of the cellulose to remove residual acids.

Cellulose nitrate was plasticized with camphor, which can vaporize over time, and other additives may migrate to the surface, leaving the celluloid cracked and brittle. UV light, high temperatures, and moisture can cause cellulose nitrate to convert its nitrogen oxides to nitric and nitrous acids. This acidic reaction can severely degrade the object and surrounding ones. It is also autocatalytic, meaning that once the acidic reaction starts, the degradation can be slowed under the right storage conditions but not stopped [FIG. 10]. Moreover, the acidic gases formed during decomposition are inflammable, and under some circumstances, combust at temperatures as low as 106 degrees Fahrenheit. Burning even in the absence of oxygen and igniting released explosive gases, these fires can be devastating to both human life and cultural heritage. As cellulose nitrate was the most popular material for film during the earlier part of the twentieth century, it created great risks as running highly nitrated or degraded film through a projector with high-temperature lamps was a significant hazard for projectionists and theatergoers. While manufacturing, chemical instability, and environmental factors are the most likely cause of cellulose nitrate decomposition, famed movie director and actor Orson Welles put the onus onto the preservation professional to mitigate this: "Film has a personality, and that personality is self-destructive. The job of the archivist is to anticipate what the film may do—and prevent it."

Prevention of cellulose nitrate combustion came in the form of "safety film," or film made of cellulose triacetate, in the 1920s, but cellulose nitrate continued to be used in many forms, including film, for years to come. Cellulose triacetate and acetate film have their own sets of preservation issues, however. Like cellulose nitrate, the cellulose polymer chain is modified, but through acetylation rather than nitration. Unfortunately, the plasticizer, triphenyl phosphate, that was often used in cellulose acetate as a flame retardant would migrate to the surface, leading to shrinkage, warping, delamination, and brittleness. Deacetylation, the replacement of acetate groups with hydroxyl groups, would occur as a result of exposure to moisture, creating acetic acid. This degradation is called vinegar syndrome because the acetic acid can smell pungently of vinegar. The reaction is autocatalytic, and the release of acidic gases can cause the corrosion of materials and objects in proximity to the degrading cellulose acetate.

Cellulose acetate and nitrate were used not just for film and photography but household products and sculpture too. Artists like Naum Gabo,

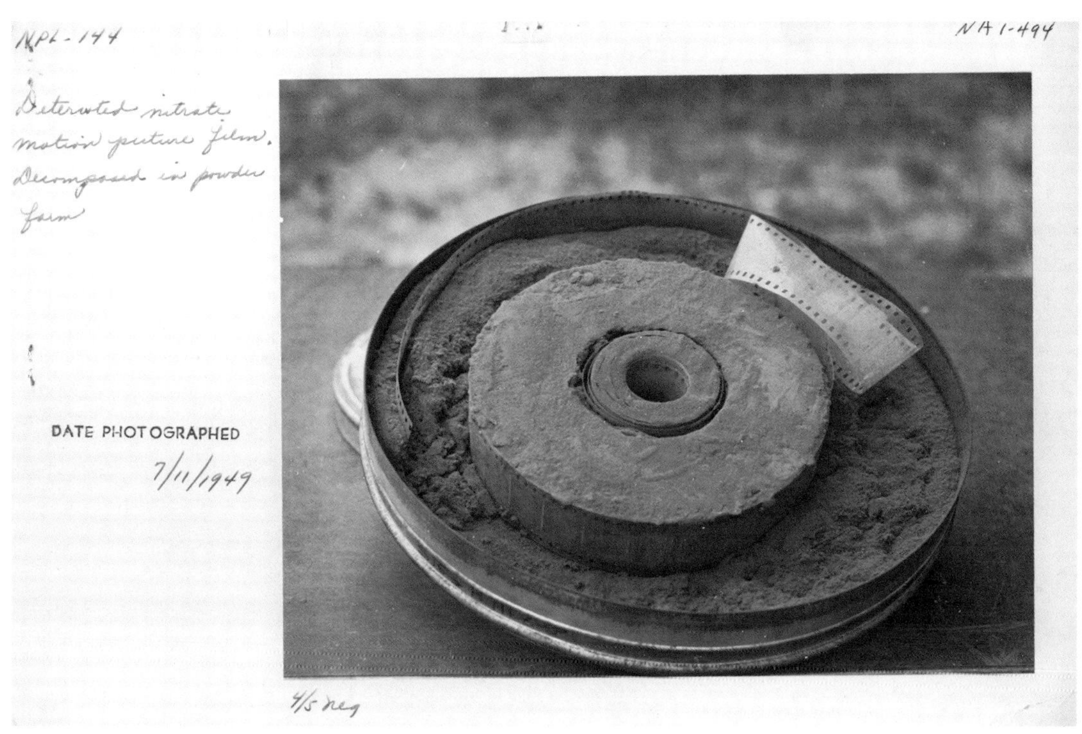

[FIG. 10]
Photograph of deteriorated nitrate motion picture film decomposed in powder form, 1949. National Archives, Washington, DC.

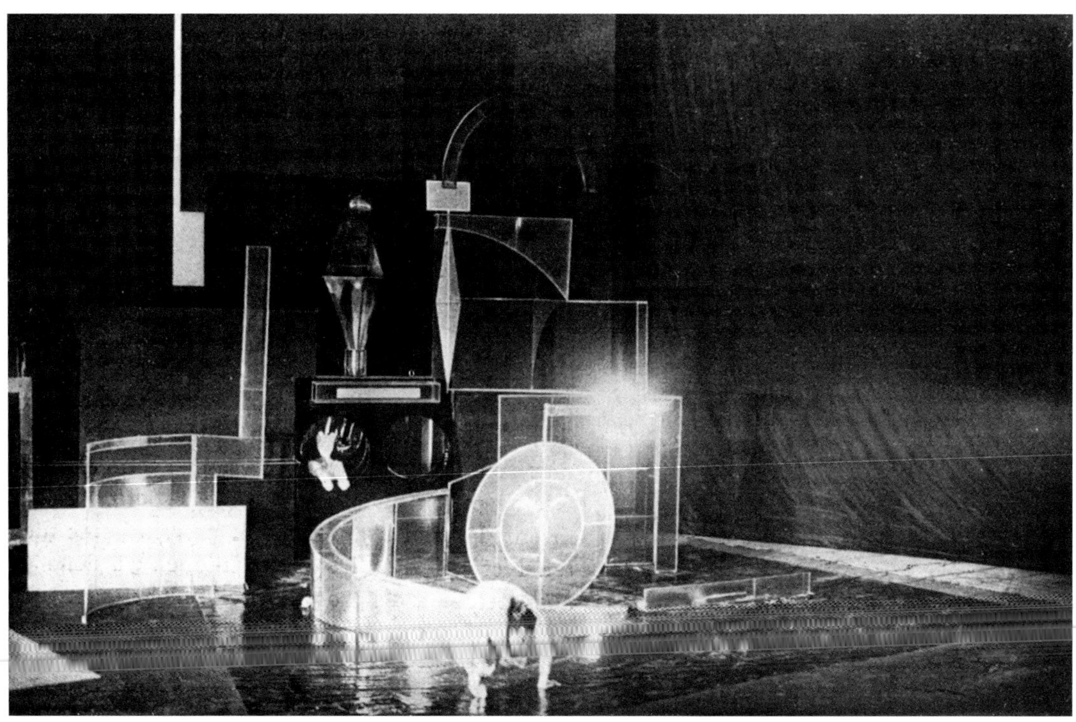

[FIG. 11]
Naum Gabo, photograph of the stage set of *La Chatte*, ca. 1927. Tate.

as discussed by Elena Torok in this volume, and Antoine Pevsner were early adopters of cellulose nitrate and acetate sheets for use in sculpture along with set and costume design, as seen in the Ballets Russes's 1927 production of *La Chatte*, for which Gabo exploited the malleability and transparency of the cellulose sheets for the set design [FIG. 11]. As described by W.A. Propert in *The Russian Ballet, 1921–1929*, the plastic structures "rounded off the stage picture beautifully, and the lights as they flickered on them had made the whole place shine with a quicksilver radiance."

PVC as well as PUR foam may not be as hazardous to a museum's health as the cellulose-based plastics, but they are difficult to preserve as objects because of their degradation mechanisms. They are also so prolific as materials that they are frequently found as components or the predominant medium of museum objects. PVC is used in modern and contemporary sculptures as well as costume and textile collections. Artist Kerry James Marshall uses rigid PVC panels as the support for his paintings.

In addition, PVC is used in countless commonplace objects as it is cheap. Flexible PVC has significant amounts of plasticizer added to the formulation, which increases its flexibility and moldability. Yet the plasticizers easily migrate, leaving a sticky or glossy surface, which attracts and traps dirt and dust, and also leads to the object sticking to itself. The migration leaves a brittle polymer substrate, which is susceptible to cracking. This can be problematic for large inflatable objects made of PVC such as *Trump Baby* designed by Matt Bonner, a piece now in the collection of the Museum of London, or Sanford Bigger's *Laocoön (Fatal Bert)* in the Brooklyn Museum of Art. It is neither ideal to keep flexible PVC stored folded on itself nor keep a large object completely inflated while not on display. Tailored long-term conservation and storage plans are required for their ongoing preservation.

PUR foam is found in furniture collections as well as sculptures, costumes, and textiles. It can be the stuffing of a Knoll chair or padding of the tongue of a shoe. But it can also discolor, turn brittle, and crumble. PUR esters are susceptible to hydrolysis and so humid conditions can be detrimental, while PUR ethers are particularly sensitive to degradation from light and UV radiation.

Degraded plastics and polymers are a problem for conservators in another way: dust. When dust lands on museum collection objects, it can obscure their appearance and in some instances damage surfaces. Research has shown that plastics, or microplastics, in the form of tiny particles and fibers have been found in huge quantities even in the most remote of areas. The study approximated that more than 1,000 tons of plastic from the atmosphere are delivered each year to protected areas in the western United States, including national parks and wilderness areas.

This is equivalent to approximately 120 to 300 million plastic water bottles. Plastics now comprise the very dust that conservators seek to remove from museum collections.

Plastics as Solutions?

The history of semisynthetic and synthetic plastics aligns closely with that of the Western field of the conservation of cultural heritage, with museums such as the State Museums of Berlin, Germany, and Metropolitan Museum of Art in New York City establishing positions and protocols in the 1870s and 1880s. As conservators fix and stabilize broken objects, one of the main tools of the trade is adhesives. In the early part of the twentieth century, despite the development of cellulose and Bakelite, most adhesives being used were still of animal or vegetal origin. These included animal and fish glues, shellacs, and natural rubbers, which have been used for millennia. The choices of the adhesives, however, used in the conservation field expanded from the traditional natural glues to synthetic adhesives with the discovery and development of semisynthetic and synthetic polymers. As Irving Skeist and Jerry Miron write in the "History of Adhesives," "Adhesives are the diplomats of the polymer world. They exist for the purpose of bringing other materials together."

One of the most diplomatic or useful of the adhesives may be Elmer's Glue-All, readily found in homes and schools across the United States. The modern Elmer's white glue is an aqueous emulsion of the polymers polyvinyl acetate, polyvinyl alcohol, and propylene glycol. Yet the bovine who has graced the label since 1951 is a reminder of Elmer's more natural predecessors. In the 1920s, the Borden Company, which started in the dairy industry as the New York Condensed Milk Company in the mid-nineteenth century, began to expand its business. In 1929, the company acquired the Casein Company of America, which made an adhesive of casein, a by-product of skim milk. This acquisition became the foundation of Borden's interest in the chemical industry, which was solidified in 1947 when Borden merged Casein with another acquired company, Durite Plastics, to form the Borden Chemical Division. In 1947, Borden introduced the synthetic white glue Cascorez, a moniker that may allude to casein and resin. Borden developed the mascot Elsie the Cow in 1936 for its dairy products, followed by her husband, Elmer. In 1951, Cascorez was rebranded as Elmer's with the face of Elmer the bull on the label. Elmer became the mascot for Borden's chemical division, a reminder of the products' origins in dairy products and casein. Elmer's Glue-All has been used as a repair adhesive throughout its history, although it was preceded by other semisynthetic and synthetic products. Like the historic development of plastic products, the first semisynthetic adhesive used in conservation was cellulose nitrate.

An important application for celluloid lacquer is that of the conservation of metal. In a 1907 publication about cellulose and cellulose products, the author notes that if collections of arms and armor are "provided with a scarcely perceptible layer of celluloid, they retain their polish and need only from time to time be freed from dust by wiping with a woolen cloth, to be kept in perfect condition." *The Third Report of the British Museum on the Cleaning and Restoration of Museum Exhibits* describes the use of cellulose acetate and celluloid (cellulose nitrate) as "amongst the most useful varnishes and cements." It suggests the use of old photographic film as a source for the celluloid but instructs that the gelatin coating must be removed to avoid gelatin in the celluloid solution. The British Museum used celluloid varnish or lacquer to consolidate the surfaces of clay Mesopotamian cuneiform tablets. This surface allowed these objects to remain intact (and not disintegrate) so that they could be soaked in water to remove damaging salts within their structures that would cause further deterioration to the tablets.

By the late 1920s, however, cellulose nitrate lacquers and adhesives had been shown to have poor aging properties, becoming weak, acidic, and brittle as well as yellowing. Nevertheless, by then fully synthetic adhesives and resins (including phenol-formaldehyde and polymerized vinyl products), urea derivatives, and ketone and sulfur phenol resins were available, as noted in the 1923 publication *Synthetic Resins and Their Plastics*. The possibility that research chemists could improve on natural resins, which had become scarce, is highlighted, as synthetic resins entering the market herald the "activity of research chemists who already have demonstrated that it is possible not merely to equal but in many cases to outdo nature by producing synthetically resins having properties greatly desired but not heretofore found in natural products."

The Straus Center for Conservation and Technical Studies (founded in 1928) at the Harvard Art Museums began to experiment in 1929 with the use of "Vinylite A," a polyvinyl acetate, as a replacement for the animal glues used during the removal of wall paintings from their original locations at archaeological sites. With any method, removing a wall painting from its original architecture is an invasive and inherently destructive process. It was an all-too-common approach in the early twentieth century to remove art from archaeological sites, with the intention of preservation in Western museums and private collections. In the process of removal, the paint layer is "faced" with a layer of adhesive and fabric to protect the painted surface. The experiments at the Harvard Art Museums with polyvinyl acetates for the facing process was a result of significant problems that occurred to wall paintings removed from the site of Dunhuang in China in the 1920s [FIG. 12]. The animal glue that was traditionally used froze in the cold of the cave, damaging the paintings.

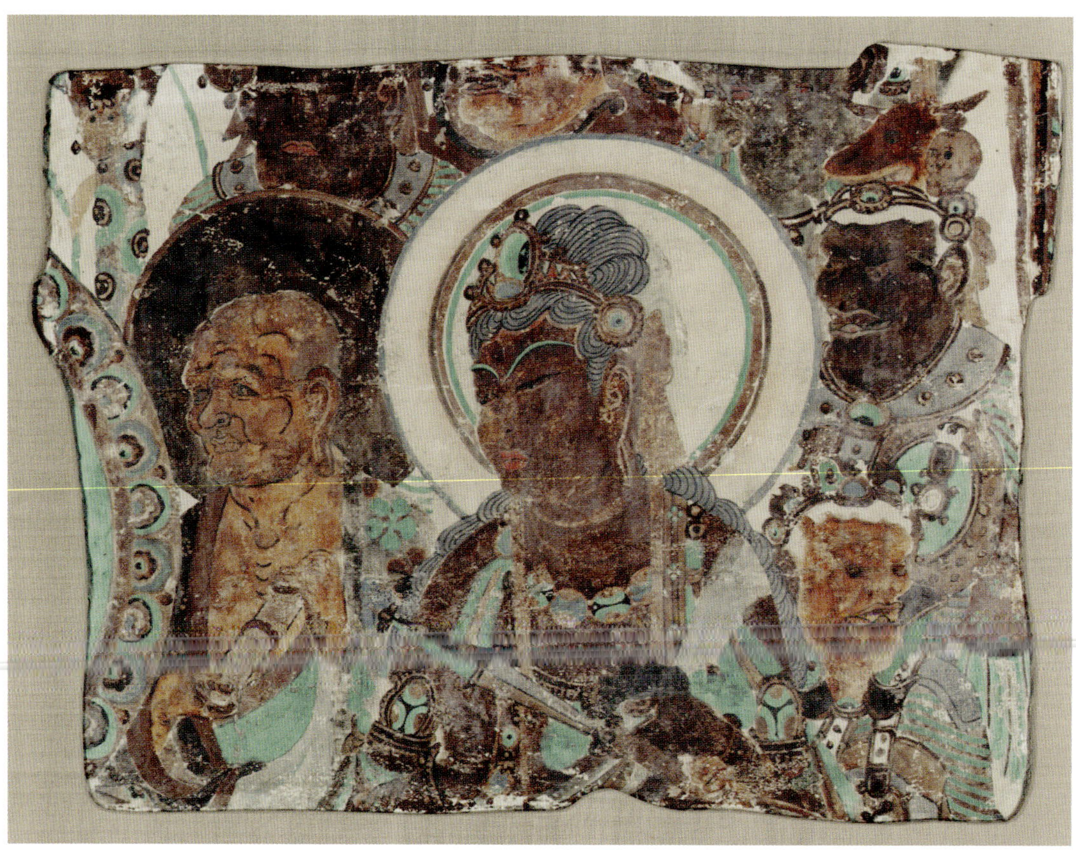

[FIG. 12]
Anonymous, bust of a bodhisattva surrounded by a monk and devas, from the south wall of Mogao Cave 320, Dunhuang, Gansu Province, early eighth century. Wall fresco removed with adhesive. Harvard Art Museums, Cambridge, MA.

Advances in adhesive technology progressed from the mid-1940s onward with the development and commercial availability of nitrile-phenolic, vinyl-phenolic, acrylic, and PURs in the 1940s; epoxies and cyanoacrylates in the 1950s; and continuing to the present day. As the field of conservation is small, adhesives are not usually developed specifically for conservation, but the field adopts adhesives from other industries as suits the needs of conservators. Adhesive selection is based on long-term stability and aging characteristics. Adhesives that are commercially available can have proprietary recipes, making it difficult to know if unidentified additives will have an unintended effect on the artwork. The adhesive should have a high degree of reversibility so that it can be solubilized and removed with relative ease in the future should retreatment be necessary. The adhesive and its application method should be sympathetic to the material on which it is applied—not causing further damage to the object.

Beyond the diplomatic adhesives, solutions with plastics come in any number of forms for the care of museum collections: polyethylene sheeting to cover objects to create a barrier from moisture and dust; Plexiglas cases for the exhibition of materials; polyolefin tubing to pad mounts to protect objects from damage while on display; polyester threads for stabilizing textiles; PET film or Mylar sleeves for enclosing photographs and archival material; synthetic coatings and varnishes to prevent corrosion and degradation; and 3D-printed replicas of objects or missing parts. It is difficult to imagine conservation work without synthetic polymers.

Marching into the Future

Plastics are now solidly part of our cultural heritage and are well represented in museum collections as well as used in the care of those collections. There is no kingdom in which plastics do not exist. The future is full of them because they will not go away, as Elizabeth Bradfield alludes to in this excerpt from the poem "Plastic: A Personal History":

Do you know
what I mean? I mean, what pearl forms
around a grain of plastic in an oyster?
Is it as beautiful? Would you wear it?
Would you buy it for your daughter
so she in turn could pass it down and
pass it down and pass it down?

To conserve museum collections is to thwart disposal, and "pass it down and pass it down and pass it down." Even plastics—the deceptively simple, banal, and beautiful.

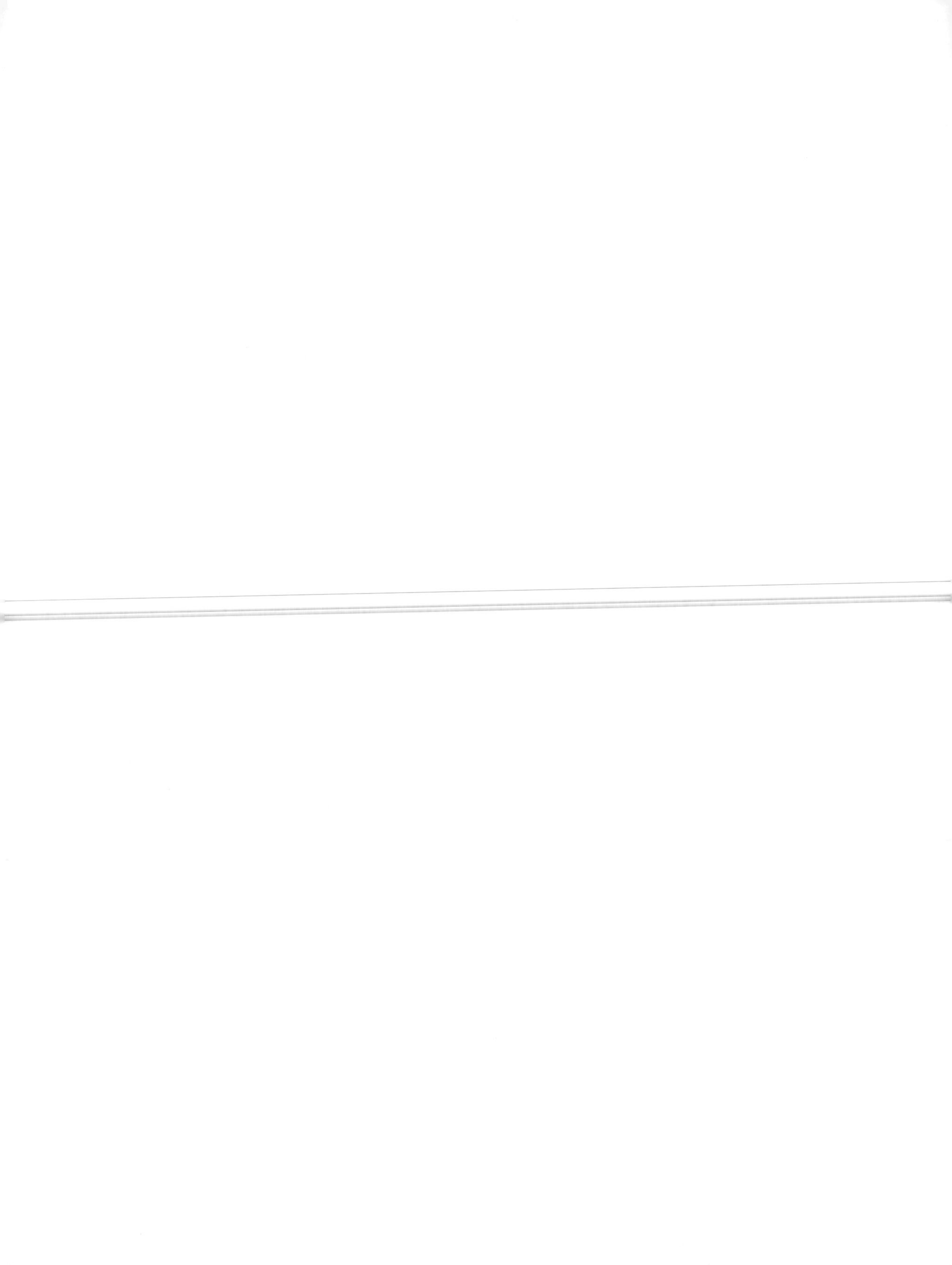

Plastic Histories of Art

David Joselit

The ostensible modesty of plastic is inseparable from its monstrosity; it is the banality of each individual container, Styrofoam brick, or sheet of plastic wrap that belies the vast scale of their accumulation as trash in oceans, landfills, and city streets. To account for the aesthetics of plastic, it is necessary to confront this paradox. For if there is any reason to interrogate plastic as an artistic material—as this book sets out to do—it is to recognize the vertigo induced by its invisible ubiquity—what we might call the plastic sublime. Plastic functions in the background of our lives as a type of toxic waste marking what author Rob Nixon has called the "slow violence" of the consumer society that metastasized after World War II, but also as a powerful form of protection that enables the hygienic storage of food and safe operation of contemporary medicine. In other words, plastic mitigates the very risk that its toxicity exacerbates. As a material used to create sterile, one-use medical products ranging from syringes to intravenous bags, plastic assists privileged citizens of the West in carving out a privatized space free from risk—a kind of neoliberal bubble. In short, plastic is characterized by its *plasticity*—by its capacity to change its form, not only as a material, but as a social agent. To comprehend its aesthetics, therefore, we must confront its capacity for metamorphosis, for it is this quality of transformation rather than any single physical characteristic that ultimately defines the aesthetics of plastic.

As I see it, three paradoxes are fundamental to the behavior of plastic, and I will enumerate each in turn.

Plastic Is Both a Material and Action, Both Singular and Universal

When one introduces a material into discussions of modern art, it is the concept of *medium* that typically prevails. Paint, marble, clay, wood, steel, or film, may all establish a one-to-one correspondence to a practice or aesthetic action such as painting, sculpture, ceramics, film, photography, or video. Plastic confounds such categorization. On the one hand, it is a component of the material substrate of many media, but unlike the relation of paint to painting, this plastic content does not rise to the significance of a regulating concept for a coherent set of aesthetic practices—that is, a *medium*. Film was initially made from celluloid, an early form of plastic; acrylic paint contains plastic, but it is unproductive, and even absurd, to discuss film or painting in terms of plastic as a determining feature of their production. On the other hand, the self-conscious use of plastic as a material in art largely dates from the period of the so-called Great Acceleration after World War II, when plastic production exploded. And yet the preponderance of this work either uses plastic opportunistically (in technological procedures that may or may not prioritize the meaning of it as a material) or "archaeologically," through appropriating and recombining discarded plastic objects made for other purposes. In some cases, there is virtue in thinking of "plastic" as a meaningful interpretative framework, but in others—such as the plastic boxes used by artist George Maciunas to hold Fluxus objects—there is not. In short, if plastic is not a medium but rather a proliferation of materials of widely varying significance to the meaning of the artwork in which they are included, how should we theorize an aesthetics of plastic? This conundrum leads to a consideration of plastic as an *adjective* as opposed to a noun. For in several languages, the quality of *plastic* denotes not an artistic material but instead the very principle of creating aesthetic form. This is why the now-obsolete phrase *plastic arts* was once in wide circulation. Here, then, is our first paradox: as a material, plastic is too peripheral to art making and too multiple in its forms to correspond to a medium. But as an action, *plastic* (or the plastic arts, or plasticity) is too general to have any analytic meaning in that it denotes the universal procedures of formation, of making form, that underlie every artistic medium and every individual artwork. One solution to this paradox is indicated by philosopher Catherine Malabou's tripartite concept of *plasticity*. As she observes in her book *What Should We Do with Our Brain?*,

> The word *plasticity* has two basic senses: it means at once the capacity to *receive form* (clay is called "plastic," for example) and the capacity to *give form* (as in the plastic arts or in plastic surgery). . . .
>
> But it must be remarked that plasticity is also the capacity to annihilate the very form it is able to receive or create. We should not forget that *plastique*, from which we get the words *plastiquage* and *plastiquer*, is an explosive substance made of nitroglycerine and nitrocellulose, capable of violent explosions.

In the definition of plastic aesthetics I will offer, it will not be enough for an artwork merely to contain elements of plastic material. Instead, a work must engage in a dynamic of formation and deformation; of receiving, giving, and annihilating form using plastic as a material that explicitly demonstrates the dynamic of plasticity. In this act, the singularity of a particular type of plastic substance included in an artwork converges with the universality of plasticity as a principle of form making. Later on, I will consider works by artists Marcel Broodthaers, Arthur Jafa, and Matthew Angelo Harrison that, in different ways, conform to this definition.

In Plastic, Deep Geological Time Encounters Instantaneity

The temporal signature of plastic is not only paradoxical but also vertiginous. It takes millions of years for organic matter to be transformed through geological pressure into oil (the primary raw material for plastic). Such duration is frankly beyond human comprehension, but the products of this prehistoric material, including single-use plastic products such as packaging and sterile medical devices, the production of which exploded after World War II, have an entirely different time frame. Plastic is a material with a useful life that is, if not instantaneous, typically short; it may enter our consciousness for as little as a few days, few hours, or few minutes. But once it leaves our hands, *as waste*, plastic returns to deep time. As a material slow or resistant to biodegradation it will remain in the earth [illegible] its landfills, oceans, and urban and rural landscapes—for hundreds of years. In short, our encounters with plastic constitute an instantaneous interlude between two eternities, of which we are largely unconscious.

In a sense, this literal carelessness, this squandering of deep time for momentary convenience, for the avoidance of risk, and to ensure temporary well-being, describes the ideological disposition that has led to the condition of the Anthropocene, understood as a geological era defined by the human intervention of fossil-fuel-generated industrial revolutions. And indeed, the term *plasticene* has appeared in publications as diverse as the *New York Times* and professional scientific papers. In one of these authored by Linsey E. Haram, James T. Carlton, Gregory M. Ruiz, and Nikolai A. Maximenko in 2020, "A Plasticene Lexicon," the plasticene is defined as follows: "an era in Earth's history, within the Anthropocene, commencing in the 1950s, marked stratigraphically in the depositional record by a new and increasing layer of plastic." Scientists have even identified a novel kind of geological material, the plastiglomerate, composed of melted plastic along with materials like sand and rock. As we will see, the collision of instantaneity and geological time has led some artists to understand plastic objects as fossils in their effort to conduct an archaeology of the present.

Plastic Is Associated with Both Life and Death

The preponderance of recent commentary on plastic has dwelled on its status as waste, especially in terms of its toxicity and virtual ubiquity, in the environment as well as human and animal bodies. Much less common is an acknowledgment of how pervasively contemporary plastic supports life. One exception is a moving 2013 essay by Jody A. Roberts, "Reflections of an Unrepentant Plastiphobe," in which the author, a science and technology studies scholar, recounts the change of heart he experienced regarding plastic after his infant daughter required intensive medical care and the very material he had assiduously avoided as a "plastiphobe" was used abundantly in her lifesaving treatment. Roberts is forced to acknowledge to himself and his readers that a binary relationship to plastic—as a material causing death versus one enabling life—is untenable. If plastic toxicity and waste contribute to illness in humans, its presence in medical procedures combats illness. What plastic does to keep us safe—from preserving food to housing intravenous fluids—is seldom recognized in academic and popular discussions of the material, leading to one of its least acknowledged paradoxes as an agent of both illness and health. Environmental scientist and critical theorist Max Liboiron, for instance, insists on the necessity of understanding plastic as part of an ecology of life, as opposed to merely a carrier of toxicity. This is particularly powerful coming from Liboiron, who directs a lab devoted to studying plastic pollution in marine life. In their 2021 book *Pollution Is Colonialism*, they write,

> If plastic-eating bacteria or fungi actually scaled to the point that they could address these plastics, they would also be eating all the plastics in bridges, airplanes, automobiles, pacemakers, and buildings. Everyday infrastructure would be crumbling around us, like a B horror movie. Plastics undergird and prop up most urban, many rural, and even bodily infrastructures. They literally support life.

If plastic has now moved deep into the infrastructures of contemporary life, far beyond the most visible forms of pollution, it has also fostered new kinds of life within broader ecosystems by, for instance, supporting bacteria colonies in the mid-ocean. This entanglement of plastic and life produces what is now sometimes called the plastisphere. As a term, *plastisphere* complements *plasticene* (there would be no plasticene without a plastisphere), but their connotations are quite distinct. While the plasticene denotes the human intervention in geological history that is more typically called the Anthropocene, and urgently calls for remediation, the plastisphere allows space for what Liboiron would call building "good relations" between plastic, the land, and human beings. What this means is to acknowledge both the risks and benefits of a particular material, learning to live with it in a way that reduces the former and enhances the latter. What is important for our purposes is that much progressive discourse tends to overlook the complexity of plastic's insinuation into our daily lives. It is too late for its eradication, but the risks of plastic pollution—which only increase with increased consumption worldwide—is nevertheless too dire to be ignored.

* * *

If plastic, then, is both a material and an action, if it opens onto two radically different temporal scales of geological time and instantaneity at once, and if it simultaneously aligns with life and death, how can one address it aesthetically? As I have already argued, an approach based on artistic medium in which one would establish a coherent historical genealogy of art defined by its utilization of plastic is inadequate to the task. The explanatory function of medium specificity within art history has been to create linear, evolutionary histories (with all kinds of delays, digressions, and "regressions") based on successive strategies for manipulating a material such as marble, paint, or video. We have accounts of European painting, for instance, that move from the evocation of perspectival depth to an acknowledgment of the materiality of paint as a viscous fluid on a flat canvas. While the familiar story of Western painting since the Renaissance—from illusionism to abstraction—has been complicated in many significant ways by several generations of art historians, *medium* remains the dominant conceptual framework for linking a material such as paint to a set of procedures like perspective or impressionism.

An aesthetics of plastic will not and cannot conform to such a medium-based account; the use of plastic as a constituent material in artworks simply does not follow such a course. Consequently, meeting the challenge of narrating an aesthetics of plastic holds the promise of offering novel interpretative frameworks that may be relevant beyond the realm of polymers. As I've already hinted, the theoretical model that I will pursue deploys theories of plasticity—those of philosophers Roland Barthes and Malabou as well as literary theorist Zakiyyah Iman Jackson—to trace a rhythm of aesthetic transformations and metamorphosis distinct from the regulated progress of a medium-based approach. Despite its brevity, Barthes's short essay on plastic, published in his influential book *Mythologies* in 1957, remains central in subsequent accounts of the material. This is because Barthes partly resolves the paradoxes I have enumerated by treating plastic not simply as a substance but also as a dynamic principle. He writes,

> So, more than a substance, plastic is the very idea of its infinite transformations; as its everyday name indicates, it is ubiquity made visible. And it is this, in fact, which makes it a miraculous substance:

> a miracle is always a sudden transformation of nature. Plastic remains impregnated throughout with this wonder: it is less a thing than the trace of a movement.

By addressing plastic as *both* a substance and "the trace of a movement," Barthes demonstrates that what I have treated as paradoxes might better be understood as *material qualities*. But this still leaves significant questions. For example, if plastic is characterized by transformation, does such transformation have a set of historical forms, or as Barthes suggests, are they *miraculous* (i.e., inherently outside history)? And second, how can we recognize the trace of a movement in a particular plastic form? For these questions, the philosophical work of Malabou is distinctly enlightening.

I have already cited Malabou's tripartite definition of plasticity—as giving, receiving, and destroying form. This third destructive capacity enables a material operation of the dialectic: one form replaces another by occluding or destroying the first. In Barthes's account of plasticity, which is extrapolated from the qualities of plastic as a modern material, the dialectic of giving and receiving form by destroying it is absent. For him, plastic is defined as a movement, but this movement is not one of negation, as in Malabou's view. Moreover, unlike that of Barthes, Malabou's theory is grounded not in the material qualities of the substance that is modern plastic but rather, recently, in the plasticity of the brain, which she interprets philosophically according to her reading of that organ's capacity to form itself through the history of an individual's experience. As she writes in *What Should We Do with Our Brain?*,

> Our brain is in part essentially *what we do with it*. . . . Plasticity, between determinism and freedom, designates all the types of transformation deployed between the closed meaning of plasticity (the definitive character of form) and its open meaning (the malleability of form). It does this to such a degree that cerebral systems today appear as self-sculpted structures.

Malabou's sculptural metaphor for the brain is hardly arbitrary, for plasticity in all of its forms retains its long-standing relationship to art as a practice of formation. But the specificity of plasticity's dialectic indicates an even more powerful relation to aesthetics. Following neurologist Antonio Damasio, Malabou reflects on how biological impulses within the body are transformed into flows of images that constitute a mind. In other words, the dialectic that defines Malabou's notion of plasticity is the transformation from a neuronal signal to a mental image: "The sculpture of the self is born from the deflagration of an original biological matrix."

She asserts that

> we must arrive at . . . a plasticity-link that is never thought of or recognized as such, allowing us to elaborate a true dialectic of the auto-constitutions of the self. This is what we must discern, as did [Sigmund] Freud in his day by analyzing the type of transformation enabling the transition from the neuronal to the psychical, the latter never being, in a certain sense, anything more than the metamorphosis of the former. If we do not think through this transformation or this plasticity, we dodge the most important question, which is that of freedom.

Though Malabou's concerns are not art historical, her dialectical account of the constitution of an image from a material base (in her case, a mental image from the body's unconscious neuronal impulses) is highly suggestive, and offers a road map for charting how, within certain art practices of the twentieth and twenty-first centuries, the material of plastic is used to demonstrate plasticity as a dynamic process of transformation.

Broodthaers's volume *Industrial Poems*, produced between 1968 and 1972, offers an excellent starting point for defining such an aesthetics. Broodthaers, a Belgian artist, is sometimes aligned with conceptual art on account of his best-known project, the *Musée d'Art Moderne, Département des Aigles* (1968–72). This complex, multipart work reframed the museum's organizational structure in at least two important ways (hence its frequent designation as "institutional critique," a variant of conceptual art that seeks to highlight the ideological underpinnings of art's exhibition). First, works of art were reduced to what might be called the "before and after" of their exhibition [FIG. 13]: the "before" marked by the presence of crates (where objects are stored when not on exhibition) and the "after" by postcards (as ex post facto souvenirs of a direct experience of an artwork typically available in museum stores). Second, Broodthaers's *Musée* scrambled curatorial taxonomies and classifications by grouping visual artifacts according to their shared iconography—for instance, eagles in various contexts ranging from wine bottle labels to insignia—rather than by style, period, or medium.

Until he was forty, Broodthaers was a poet. His initial exhibition as a visual artist in 1964 included, as a premonition of his subsequent preoccupations with the plasticity of language, a sculpture made by embedding unsold copies of his poetry volume *Pense-Bête* in plaster, thus rendering the texts both illegible, and as elements of a molded sculpture, plastic in the traditional art historical sense. Like *Pense-Bête*, the *Industrial Poems*, each of which is a vacuum-formed relief in plastic resembling public

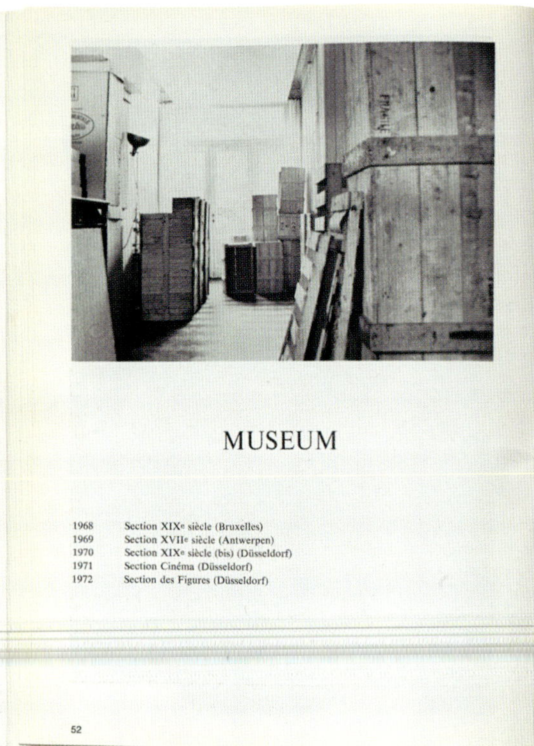

MUSEUM

1968	Section XIXe siècle (Bruxelles)
1969	Section XVIIe siècle (Antwerpen)
1970	Section XIXe siècle (bis) (Düsseldorf)
1971	Section Cinéma (Düsseldorf)
1972	Section des Figures (Düsseldorf)

52

ENZYKLOPÄDIE

(Fortsetzung)

M

Museum

Das Zeitalter der Aufklärung und Verkündung der Menschenrechte war die Epoche der ersten öffentlichen Museen. Die ersten Museumsreformen in Europa hielten mit der Entwicklung der Wissenschaften Schritt und zielten darauf hin, das enzyklopädische Museum durch spezialisierte Museen zu ersetzen — durch Museen der Naturgeschichte, der Archäologie, der mittelalterlichen und neueren Kunst usw. Es fehlte freilich nicht an Stimmen, die sich gegen diese analytische Ausstellungsweise richteten, die forderten, daß die Gegenstände in Museen durch ihre Anordnung ein dem Laien verständliches zusammenhängendes Bild eines Geschehens bieten, ihm eine „Geschichte erzählen" sollten.

Für das Museum gab es kein Palladium mehr, keinen Heiligen, keinen Christus; die Begriffe Verehrung, Ähnlichkeit, Phantasie, Schmuck oder Besitz sind mit seinen Objekten [illegible] von Dingen, die von diesen Dingen selbst verschieden sind und gerade auf dieser spezifischen Verschiedenheit ihre Daseinsberechtigung gründen: das ermöglichte ja überhaupt erst ihre Vereinigung. Ein imaginäres Museum wird die Intellektualisierung, wie sie durch die unvollständige Gegenüberstellung der Kunstwerke in den wirklichen Museen begann, zum Äußersten treiben.

Musée d'Art Moderne, Département des Aigles: Sauvez les meubles!

Mysterien

Sonne, Morgenröte, Sommer.

Licht, Blitz, Feuer,

Feuertempel, Brandpyramide, Scheiterhaufen.

Wie Adlermenschen die geflügelte Sonnenscheibe emporheben, so trägt der Adlerdämon den König, die Sonne, wie er sich nennt und anreden läßt.

Der arische Lichtgott Mithra (altindisch Mitra, griech.latein. Mithras) galt als Spender von Fruchtbarkeit, Frieden und Sieg, als Gott der Vertrages und der Freundschaft (vgl. Pan-indianischer Adlerkult).

Sein Kult verschmolz in Babylonien noch unter den Achämeniden (s. dort) mit dem des Sonnengottes Schamasch. Im letzten Drittel des 3. Jh. n. Chr. wurden die [illegible] Staatsreligion. Der Adler war Symbol der höchsten Stufe der Gotteserkenntnis. Mit ihm gelangte die Seele nach ihrer Wanderung im Tierkreis durch die Sonnenpforte zur ewigen Ruhe.

Aufstieg des Adlers, der Seele, des Gottes.

53

[FIG. 13]

Maria Glissen, view of Marcel Broodthaers's exhibition at the Musée d'Art Moderne, Brussels, 1968–69. Museum of Modern Art, New York.

signage, enacts a version of Malabou's dialectical transformation from material substance (i.e., neuronal stimulus) to mental image [FIG. 14]. But in this case, the material/representational shift occurs from poem, associated with a plastic flow of mental images, to commodity, characterized by industrial standardization and the circulation of products. As art critic and historian Benjamin H.D. Buchloh eloquently puts it in a 1987 essay on these works,

> Erasure of language in these panels results as a "natural" consequence of their fabrication in a casting process where language appears literally blinded (blind-stamped), and where it acquires the status of a relief at the cost of readability. Poetic text, artistic object, discursive classification, and institutional demarcation are all literally made "of a piece," and of one material; in their final format they are framed as mere advertisement and, in their final form, they are contained as mere object (another art commodity).

Vacuum-forming is a procedure in which the final product is both molded and itself potentially a mold. Broodthaers's use of this material is consistent with his long-standing interest in the mold as a mobile figure, crossing even into the animal world, as in his several sculptures composed of mussel shells tumbling out of deep pots. As a masculine noun, *moule* in French means "mold"; as a feminine noun, it denotes the mussel, and so these sculptures of Broodthaers engage in an elaborate pun that enacts a serious demonstration of plasticity's dialectic. In these works, mussel shells refuse to conform to the "mold" of the cooking vessel they rise out of, in what Malabou might understand as a form of destructive plasticity by which the mold (giver of form) refuses to be molded (take form), resulting dialectically in a sculpture. There is also the interesting dimension here of a gender transformation within the "same" word, from the masculine *moule* (mold) to the feminine (mussel). As Buchloh demonstrates, a different dialectic occurs in the *Industrial Poems* through the elision of the effect of molding and being molded in vacuum-forming: a transformation from poem to "instrumentalized language such as advertising," which is indicated by the adjective *industrial* of Broodthaers's title. But as Buchloh and other commentators have noticed, the *Industrial Poems* create a further, and perhaps even more consequential, dialectic between a surface and its inscription: there is no material difference between letters and "page" in the *Industrial Poems*—language is seamlessly embedded, not separated, from its material substrate. In some of the *Industrial Poems* including *Téléphone* (1968), there is a further dialectic established between a human ego and its status as information. The plaque is organized in four registers. The top two

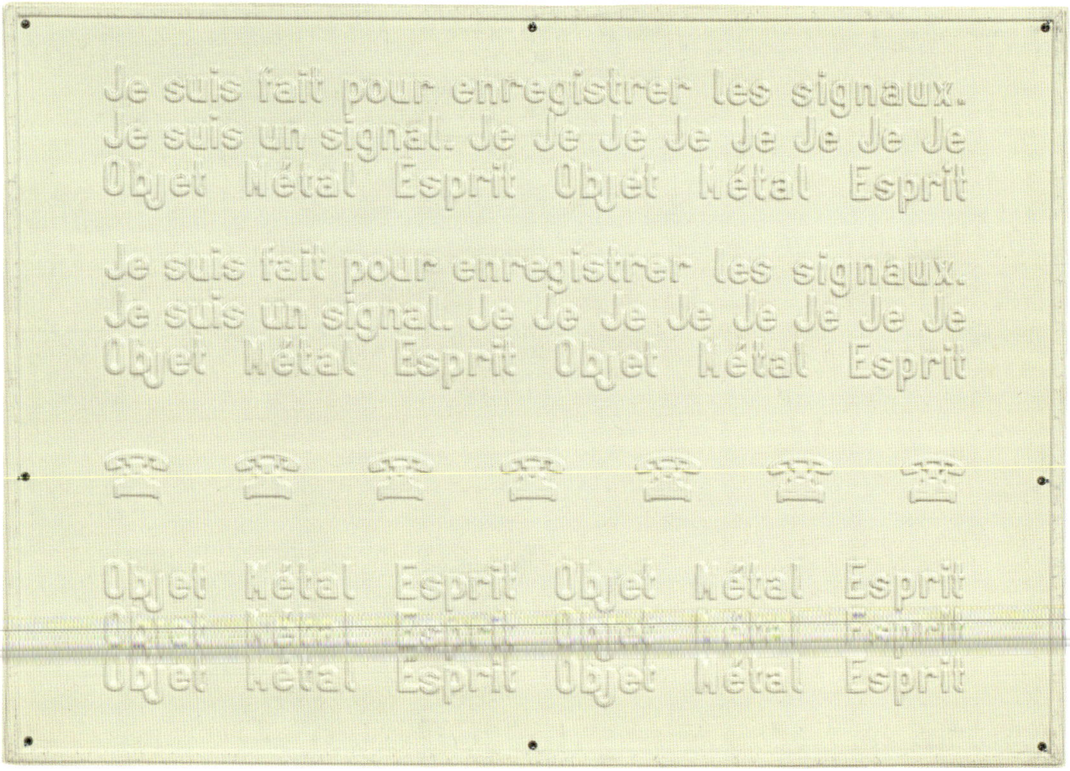

[FIG. 14]
Marcel Broodthaers, *Telephone* (*Téléphone*), 1968. Painted vacuum-formed plastic plate, 32.18 × 46.75 × 0.18 in. (81.8 × 118.7 × 0.5 cm). Museum of Modern Art, New York.

repeat the following text:

> *Je suis fait pour enregistrer les signaux.*
> *Je suis un signal. Je Je Je Je Je Je Je Je*
> *Objet Métal Esprit Objet Métal Esprit*
> *[I am made to record signals.*
> *I am a signal. I I I I I I I I*
> *Object Metal Spirit Object Metal Spirit [Mind; Sense]*

The register below this text includes seven icons for the telephone, and below that are three lines each that repeat the text "Objet Métal Esprit Objet Métal Esprit." Much could be said of this complex work, but two observations will suffice. First, there is a split within the "I" [*Je*] of the sign's text that is simultaneously an active person speaking and a passive recording device, and second, there is the coexistence, and perhaps even the elision, of the visual icon for a phone and its anthropomorphic instantiation in the pronoun *I*. In the *Industrial Poems*, plastic as a material is deployed to evoke the plasticity of the modern self as a subject of mass media who can never quite decide if they are the author of their thoughts or merely the receiver of them.

One could say, then, that nearly a quarter century before Malabou developed her account of plasticity, Broodthaers was much less sanguine about its capacity for "freedom." Such skepticism is taken much further by Jackson, who argues that plasticity was a tool used to enslave African Americans by forcibly reforming their bodies and minds. Contributing to a vibrant literature of historical and theoretical reassessments of the institution of slavery along with its afterlives, Jackson maintains that it was the plasticity of the body of the enslaved—their forced manipulation into various modes of non-, anti-, or subhumanity that both enabled slavery and the white supremacist system that undergirds it. As she powerfully asserts in her 2020 book *Becoming Human: Matter and Meaning in an Antiblack World*,

> The black(ened) are, therefore, defined as plastic: impressionable, stretchable, and misshapen to the point that the mind may not survive—it potentially goes wild. We are well beyond alienation, exploitation, subjection, domestication, and even animalization; we can only describe such transmogrification as a form of engineering. *Slavery's technologies were not the denial of humanity but the plasticization of humanity.*

Unlike Malabou's account of brain plasticity, the mental and bodily plasticity that Jackson describes is a source of excruciating unfreedom,

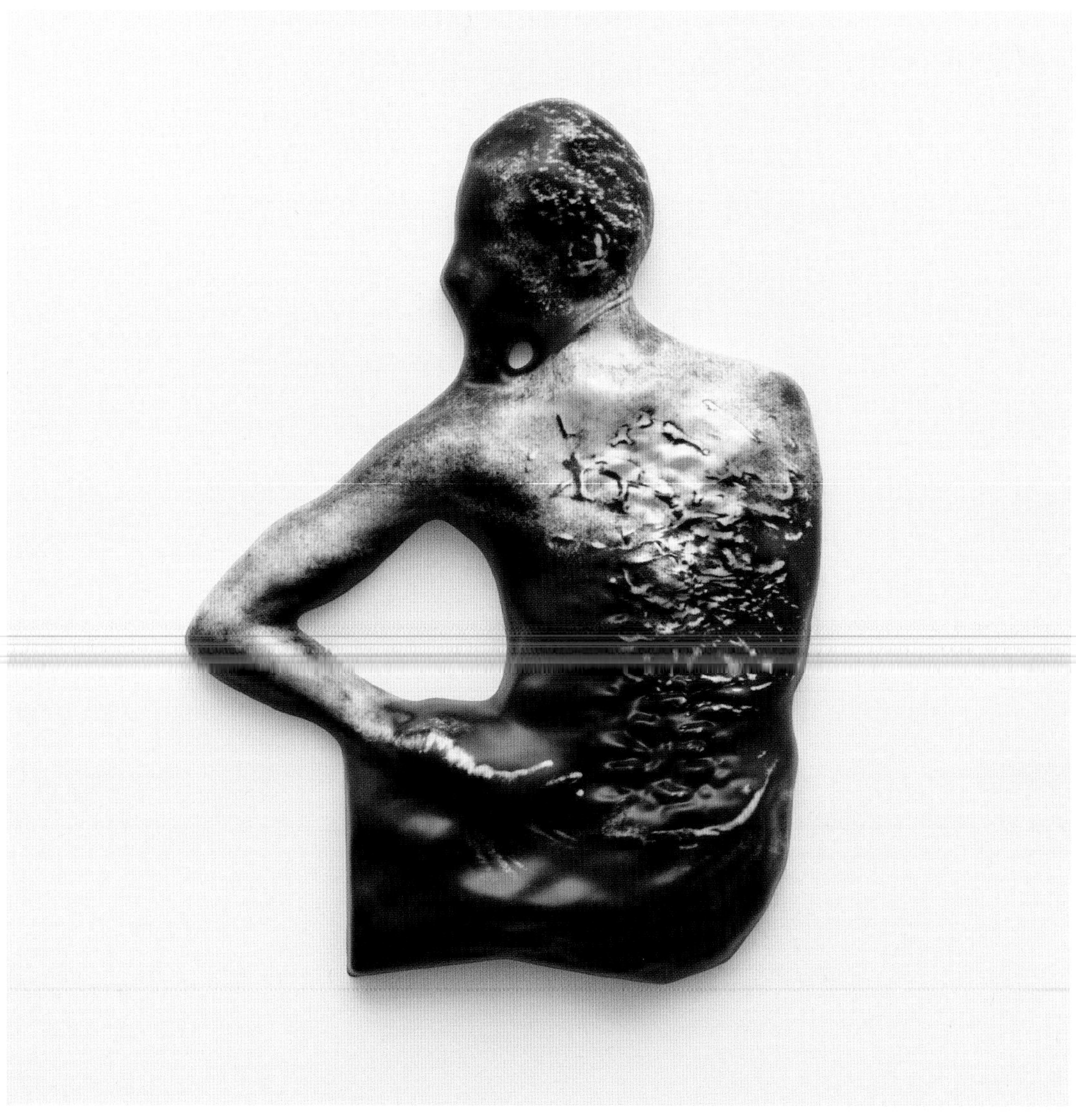

[FIG. 15]
Arthur Jafa, *Ex-Slave Gordon*, 2017. Vacuum-formed plastic, 57 × 44 × 9 in. (144.8 × 111.8 × 22.9 cm).

moving well beyond the constraints indicated in the dialectical metamorphosis Broodthaers enacts from poem to industrial sign, or from enunciation to enunciator (or author of speech to mere registrant of it) to suggest that the capacity to remold the bodies of others, both psychologically and physically, is a form of terrifying power. In fact, it is *the* central power of modern eugenics.

Two works by African American artists explore the excruciation of plasticity that Jackson identifies. In 2017, Jafa created *Ex-Slave Gordon*, a vacuum-formed plastic relief derived from a famous photograph of an enslaved man, known as Gordon, that was widely disseminated during the US Civil War [FIG. 15]. In the source image, Gordon is photographed seated, seen from behind in three-quarters' view, presenting his back scored with scars from whippings. Part of the power of this image arises from the contradiction between Gordon's elegant gesture of gripping his waist with his left hand in an almost classical pose and the evidence of violence perpetrated on his body. In Jafa's work, the source photograph is expanded in scale and shaped as a silhouette, but most important, it is rendered in relief, nearly life-size, causing Gordon's ruined back to confront the viewer more viscerally. Gordon's scars are the site of slavery's practice of plasticity—the index of physical torture meant to mold and manipulate the subjectivity of the enslaved. Moreover, the grain of the enlarged phototransfer in the rendering of Gordon's face comes to resemble the striation of his assaulted back, suggesting that it is not only the whip but also the camera that can wound. Following an observation by visual culture theorist Tina Campt regarding some of Jafa's films, I see his plastic relief as, in part, giving body to a photograph that possessed great power in its circulation. I am referring to what in her 2021 book *A Black Gaze: Artists Changing How We See* Campt calls Jafa's "still-moving images." She writes "of *still-moving-images*—images that refuse the formal opposition between photography and film. They are images that vibrate in place and in suspended motion both with and without a shift of position in time or space." For Campt, a formal level of plasticity exists in the dialectic between still and moving image, while the transformation of Gordon's photographic image into a sculpture gives his body a monumentality that it was denied under conditions of enslavement.

Harrison takes up the question of the moving versus the still in quite a different way. In several of his elegant sculptures, African artifacts are *stilled*—virtually imprisoned in blocks of acrylic [FIG. 16]. Here plastic, as a material, exhibits its capacity to immobilize an image in matter. African heritage seems to be drowning or suffocating in transparent plastic, and yet its transparency nonetheless promises visual access. The dialectic here lies in a simultaneous act of preservation and immobilization. And indeed, part of the motivation of recent efforts to repatriate African cultural

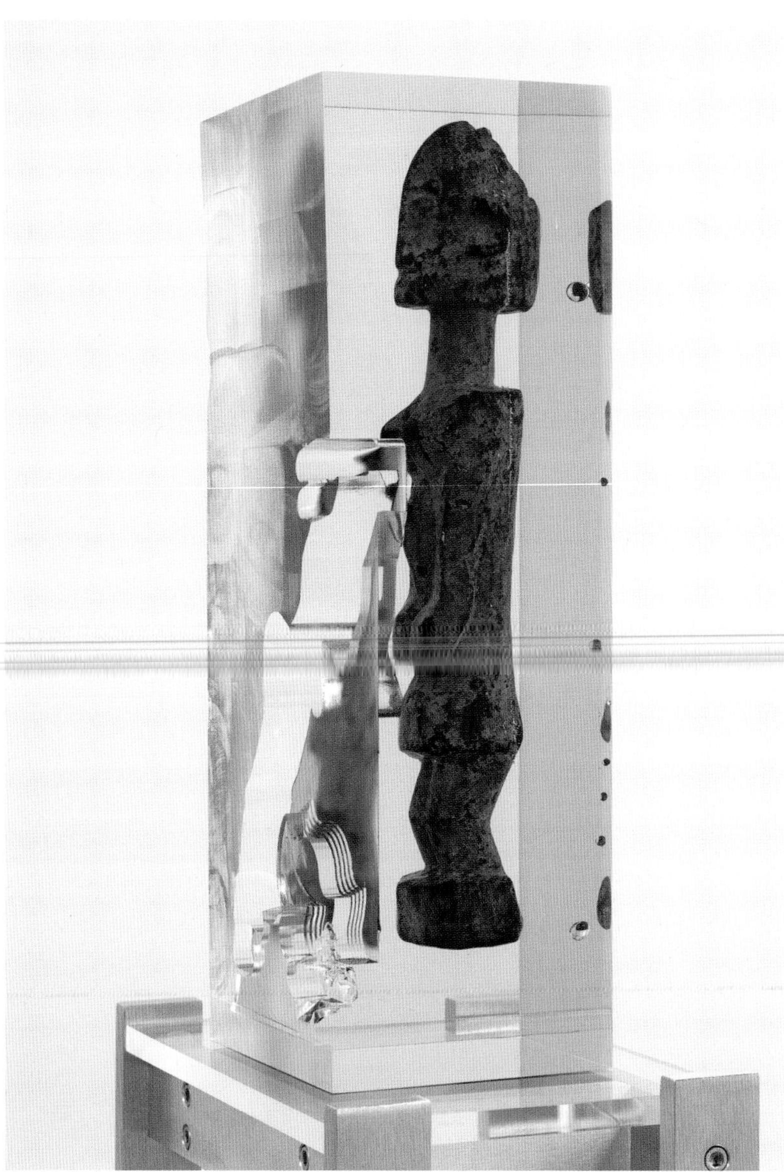

[FIG. 16]
Matthew Angelo Harrison, *Lanced Below*, 2022. Dogon wood sculpture, polyurethane resin, and aluminum stand, 56 × 11 × 7 in. (142.2 × 27.9 × 17.8 cm). Jessica Silverman Gallery, San Francisco.

properties from Western museums—objects that were often violently looted from their source communities—is that the museum continues to do violence to these works through its annexation in curatorial narratives implicitly or explicitly saturated with the values of imperialism and white supremacy.

Toward an Art History of Plastic

Following Malabou, I take plasticity to be a material dynamic of formation and deformation that results in something new. But as I have already argued, Malabou's concept of plasticity is not adequate in itself for developing an art history of plastic. For this, we must identify instances where plasticity, as a material dialectic, is directed toward plastic as a particular class of materials. In short, it is not enough for an art history of plastic to chart the miscellaneous uses and frequent appearances of various polymers within works of art, nor is a philological analysis of how "plasticity" or "the plastic" has been used as a category in Western aesthetics sufficient. I will instead identify and analyze four instances where the dialectical operation of form making parallels the qualities of the material submitted to formation—when plastic is caused to exhibit its plasticity. To a certain degree, this approach does parallel the modernist account of artistic media. Jackson Pollock, for example, certainly exploited the qualities of paint to establish a dialectical form of gestural mark making in which painterly inscriptions are simultaneously asserted as a form and denied as mimetic tools of reference. But in general, the history of Western painting has been told as a succession of developments—an evolution. The four different genealogies of plastic's plasticity that I will sketch—which are a small subset of all of those we might consider—are not sequential but rather constitute a constellation of affiliated plastic events that may be contemporaneous or overlapping in time. While they are historically specific and largely limited to the West, these dialectical instances should not be taken as periods or movements. Instead, they constitute a spectrum of possibilities for metamorphosis. I aim to evoke a history that is layered, multifaceted, and not inherently evolutionary. The four operations that I describe here—from material construction to immaterial opticality, from sentient bodies to desublimated malleable stuff, from trash to artifact, and from pollutant to landscape—thus create a field of possibility as opposed to a systematic inventory.

Constructing Immateriality

In 1916 and throughout his career, Naum Gabo made use of plastic, and as Elena Torok demonstrates in her case study in this volume, the results were often materially unstable. Gabo was a peripheral figure in the Soviet avant-garde, some of whose members developed the program of constructivism, a practice in which aesthetic principles were thought to be continuous with those of factory production. In other words, art could—and should—function as an avant-garde for industry. After moving to Berlin in 1922, Gabo, along with his brother Antoine Pevsner, is credited with popularizing constructivism as an international style. The debates between different partisans of this aesthetic tendency need not concern us here (though they are of significant historical and theoretical interest). What matters in our context is Gabo's adoption of plastic as a specifically modern material (which was hard to obtain in Russia and more readily available in Germany), and his use of plastic as a malleable and transparent material to accomplish a kind of optical sublimation in which physical stuff is transposed into immaterial, optical affect.

As a constructivist, Gabo favored structure over volume in his sculpture. In his 1916 *Constructed Head No. 2*, he built an anatomy composed of intersecting planes that evoke the shape and posture of a human body, while avoiding either portraiture or the simulation of a human skeleton. In short, neither the outer skin nor the inner structure of the body was directly evoked, but some hybrid of the two: like the cubists, Gabo built a figure furrowed with deep pockets of shadow. As he frequently did over the course of his career, Gabo remade *Constructed Head No. 2* in a variety of materials. As effective as the plastic version is, it is debatable that the material adds much to the work's meaning; other versions of *Constructed Head* were just as successful. But in sculptures like *Linear Construction No. 2* that also exist in multiple versions, the use of plastic was essential to the work's significance [FIG. 17]. Its design is composed of two intersecting planes of ovoid, transparent plastic strung with nylon filament. Though colorless, the effects are complex, with the lines of filament reflecting light and the transparent armature of the sculpture allowing light to pass through. The overall shape of *Construction No. 2* is equally complex, evoking both a complex spiral, egg-like shape in its contour and diamond shape at its core. The dialectic of plasticity it accomplishes is twofold: first, as in a double helix, the interior of the work (if such a thing exists) is indistinguishable from its exterior, and second, the solid structure of plastic disappears into an optical flash, a kind of glittering rotation reminiscent of a lantern.

The dialectic of plasticity performed by Gabo's plastic sculptures indicates two prescient revelations: that under modern conditions—dominated by the immaterial communication networks of radio and subsequently television—objects can no longer be understood as bounded entities, but rather are both penetrated by the world and extend out into it, and that a construction may produce immaterial effects rather than the more conventional association of industry with solid materials like iron and steel. It would be minimalist Donald Judd in his serial works of open

[FIG. 17]
Naum Gabo, *Linear Construction No. 2*, 1949–52. Plastic and nylon threads, 44.5 × 34.25 × 34.25 in. (113 × 87 × 87 cm). Tate.

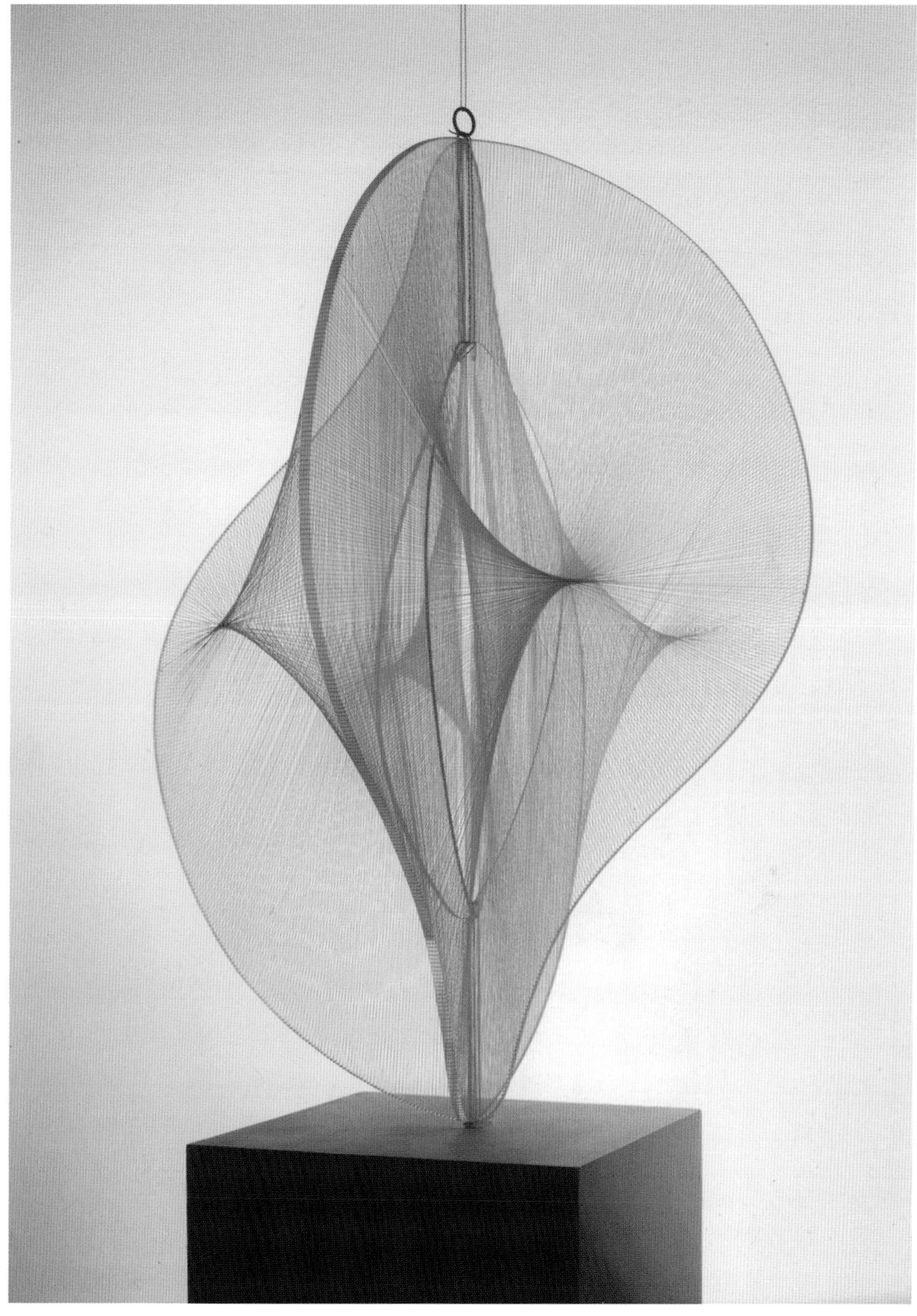

rectilinear stainless steel structures in the 1960s—whose interior and sometimes exterior elements were transparent Plexiglass sheets in bright colors such as yellow or amber—who demonstrated that the materiality associated with heavy industry is "lined" with the optical flash of the immaterial, evoking those electronic media that, along with computers, were revolutionizing the world in that decade. The open structure of Judd's parallelepipeds confounds inside and outside without dispensing with the hard edges associated with factory production. He doesn't pose an opposition between material and immaterial labor but rather reveals their thorough imbrication in one another.

Freedom versus Containment

Even though the term *plastic* has come to represent the epitome of artificiality, plasticity, as Malabou has argued, has an organic quality: it suggests a kind of transformation or metamorphosis that bodies experience when neuronal impulses are transformed into mental images, or more humbly, when nourishment is made into the building blocks of life, on the one hand, and waste products, on the other. Two complementary genealogies of bodily plasticity have animated the work of several prominent women artists since the mid-twentieth century. In one, the form of the vessel is endowed with qualities appearing but not attaining the human, and in the other, matter's malleability is performed in acts of both scatological desublimation and implied physical violence. The most prominent sculptural instance of vessels behaving as lively, even fleshly, envelopes during the mid-twentieth century is Eva Hesse's 1968 *Repetition Nineteen III*, whose nineteen components—cylindrical forms, open at the top and grouped together on the floor like a complex network of cells—are made of fiberglass and polyester resin. While reminiscent of the serial or permutational procedures of Hesse's friend Sol LeWitt, or Judd's stacks or sequences of open parallelepipeds, each of the elements in *Repetition Nineteen* is unique, each endowed with its own posture, sometimes appearing to lean toward one of its sisters rather than combining through a set of instructions as in LeWitt's work or geometric ordering as in Judd's. Most strikingly in relation to their liveliness is how the resin and fiberglass surfaces hold light, giving it the warmth and texture of flesh. There is nothing mimetically human or animal about these forms, and yet they give an impression of animation. As Caroline Fowler elaborates in her introduction, Hesse was acutely attuned to the fragility of her sculptures' "bodies"; she was aware that they could decay or even die.

If Hesse transforms the minimalist box into a biomorphic cell of aggregated cylinders, Anicka Yi, an artist of a younger generation who has been active since the late 2000s, creates what might be understood as sculptural body bags in which forms of nonhuman life are contained in plastic structures or vessels. Yi frequently uses scent in her work as well as organic materials such as bread dough and bacteria—materials that literally change their physical position and/or state in the course of an exhibition. In her 2014 work *Le Pain Symbiotique*, bread dough covered the lower surfaces of a vinyl inflatable pod, also populated with a series of sculptures in glycerine soap. In her major 2015 exhibition for the Kitchen in New York City, *You Can Call Me F*, vinyl quarantine tents were used to house the artist's own concoctions of airborne "contaminants" formed from industrial and organic ingredients. Like many of Yi's works, these tents allude to the fears of contagion that plastic is habitually deployed to protect against. In her spectacular exhibition in the Tate Modern's Turbine Hall, Yi created what she has referred to as an "aquarium of machines" consisting of two types of biologized vehicles called aerobes, small dirigibles equipped with a thermal camera positioning system and artificial intelligence. These uncanny airborne objects, which appeared as hybrids of sea creatures and mushrooms, emitted scents evoking what Yi understands as an invisible but nonetheless perceptible sculptural field—a kind of olfactory medium in which the aerobes circulated. Yi's work is clearly not "about" plastic; she deploys the material opportunistically as it is used in laboratories to contain pathogens or by engineers as a lightweight but resilient material. Her plastic enclosures suggest temporary shelter, medical emergencies, or new hybrid technologies. And yet I believe her work belongs in a discussion of plasticity's dialectic because it is precisely these plastic containment devices that separate an amorphous expanse of living material (dough, bacteria, and scent) into discernible units that lack the autonomy and authority of persons. Plastic here is mobilized as a technology that can parse forms of life beyond the human.

In some ways, Lynda Benglis's polyurethane foam sculptures of the late 1960s approach the problem of containment via plastic from an opposing perspective to that of Hesse and Yi. Benglis negates the vessel's capacity to hold viscous or unstable materials together by allowing plastic components to behave as they will in the absence of physical barriers or obstruction [FIG. 18]. This is apparent in her sculptures consisting of extravagant floor-bound mounds: tinted organic "spills" that recall viscera, vomit, or industrial waste. There is no containment here—no cylinders or hygienic tents—just the logic of gravity to which all matter on earth is subject. In the work of a contemporary artist like Anna Uddenberg, the implicit violence of such unruly plasticity is explicitly enacted on feminine bodies. In a set of procedures simultaneously suggesting coercion and perverse pleasures, Uddenberg's recent sculptures such as *FOCUS #2 (Pussy Padding)* done in 2018 are convoluted mannequins composed in part of acrylic, resin, synthetic hair, and vinyl film. In this and related works, feminine bodies are twisted into erotic furniture, presumably

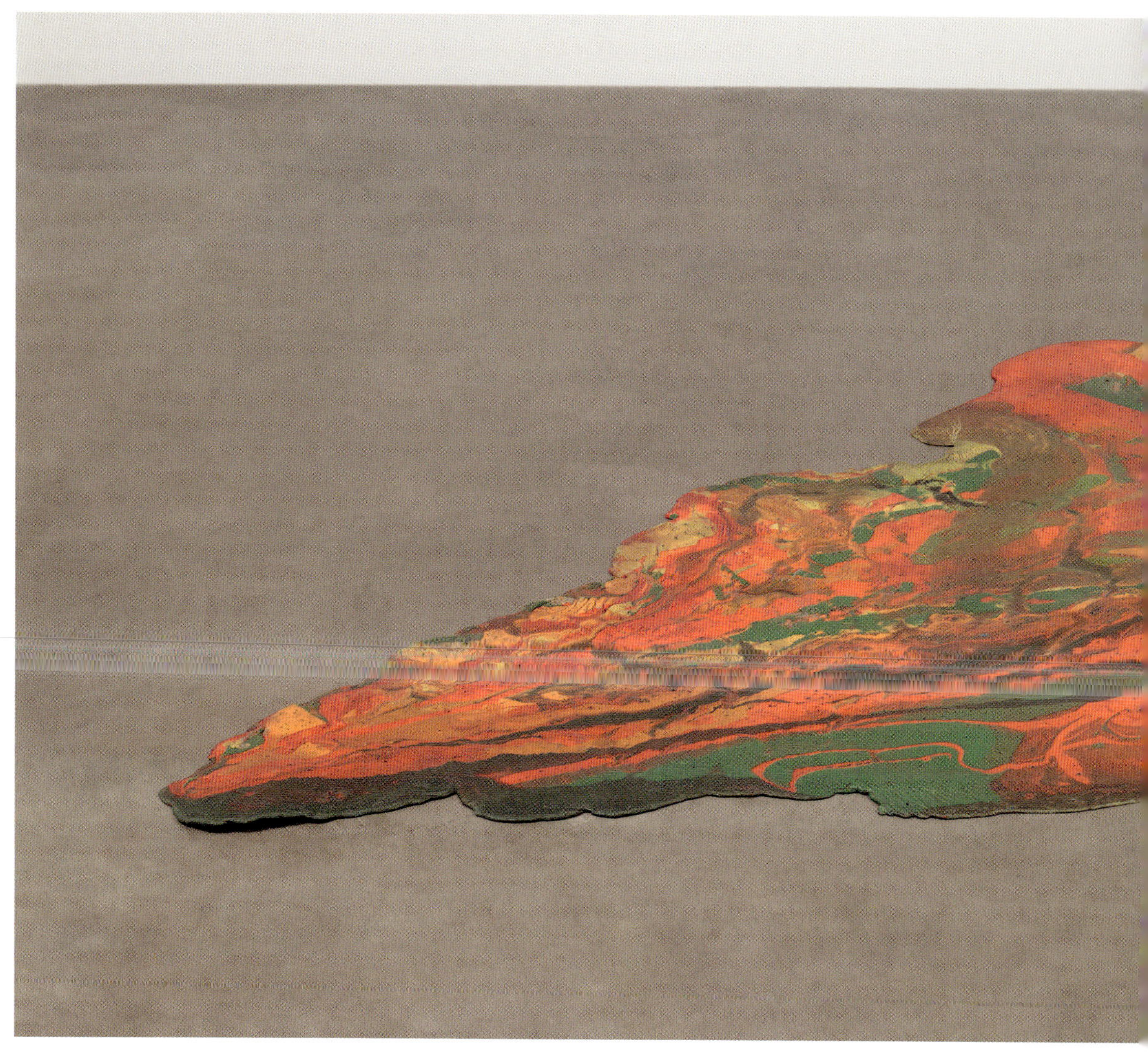

[FIG. 18]
Lynda Benglis, *Blatt*, 1969. Day-Glo pigment and poured latex, 128 × 103 in. (325.1 × 261.6 cm). Museum of Modern Art, New York. Gift of the Fuhrman Family Foundation through the Modern Women's Fund.

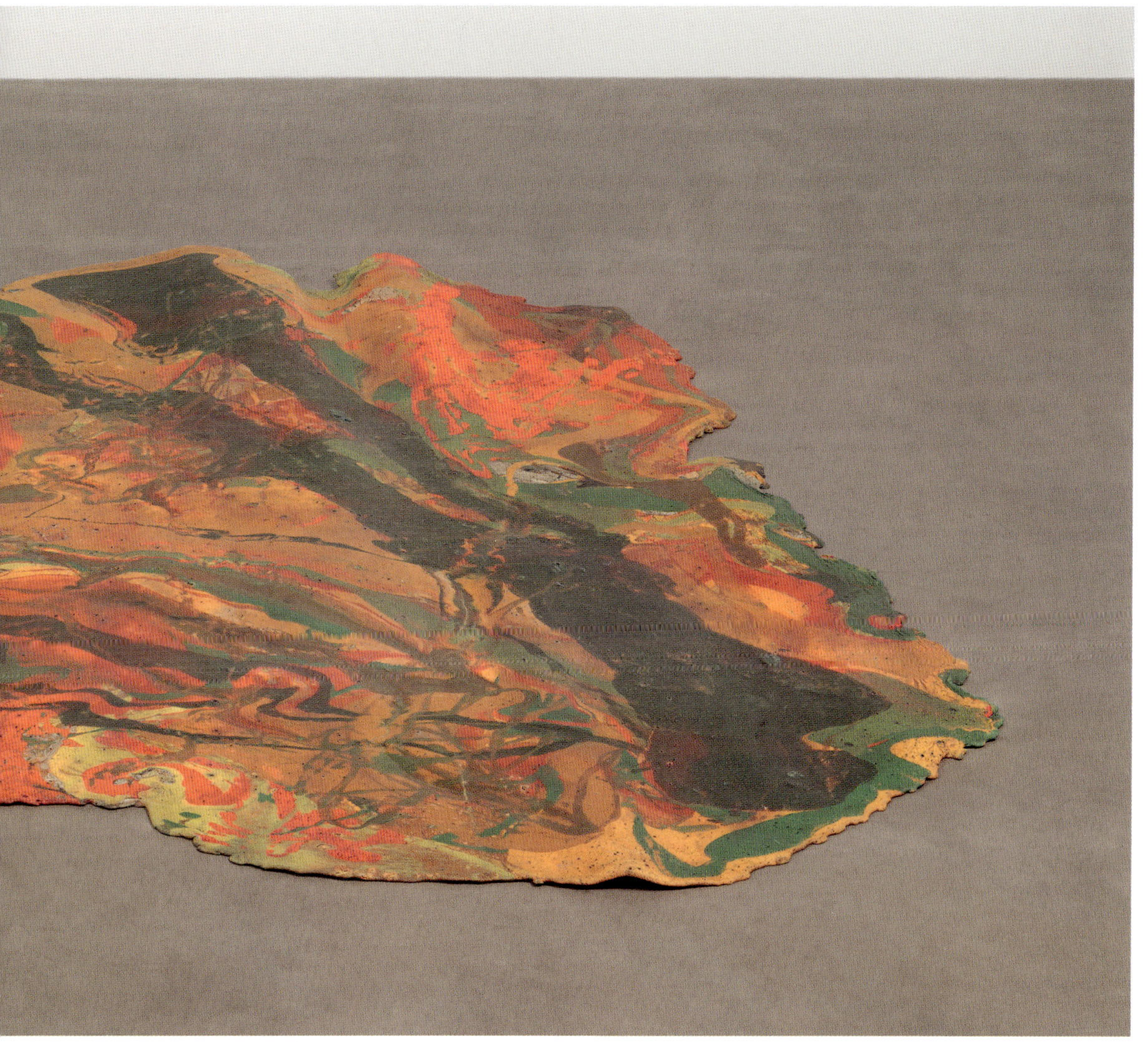

available for violation. A paradox that distinguishes Uddenberg's practice of plasticity from that of Benglis is that malleability is not aligned with freedom or desublimation but rather a coercive, even carceral form of manipulation. As this grouping of four artists of two generations demonstrates, the dialectic of material freedom versus containment does not have a stable set of political anchor points. Plasticity may be put in the service of both life's freedom and its containment.

Garbage or Artifact

William Rathje, founder of the Garbage Project in Tucson, Arizona, and whose mission was to perform an analysis of the present by charting the patterns in what people throw away, is explicit about the connection between archaeology and rubbish. He writes in his 2001 book *Rubbish! The Archaeology of Garbage*, coauthored with Cullen Murphy, "It is not entirely fanciful to define archaeology as the discipline that tries to understand old garbage, and to learn from that garbage something about ancient societies and ancient behaviors." And indeed as Rathje notes, it is only time that distinguishes garbage from museum artifacts; once it is excavated, ancient refuse may yield museum pieces. It is not surprising that in 1973, amid a mid-twentieth-century explosion of garbage in the Western world including vast amounts of plastic products and packaging, Rathje would recognize the urgency of conducting an archaeology—or garbology—of the present. He was not alone in this; during the same period, artists too were exploring the dialectic between rubbish and artifact using plastic in diverse ways to articulate their intersection. French-born US artist Arman is known for his accumulations of objects, often consisting of garbage. His aggregations of refuse are presented in at least two ways: displayed in transparent plastic containers as in *Jim Dine's Poubelle* of 1961 or "fossilized" in a resin block [FIG. 19], also encased in a plastic container, as in *Poubelle Organique* of 1971. It is ironic that Arman uses a material notorious for its contribution to the pollution problems of the twentieth century as a hygienic preservative for refuse rather than as garbage itself. Here plastic becomes a medium of conservation; it is precisely what turns rubbish into artifact. While reminiscent of Harrison's sculptural encasements of African artifacts, the reigning ethos in Arman's work is one of preservation versus suffocation—or perhaps the two always go together, but in some cases, the components collected are elevated in status, and in others, they are deprived of their meaning.

It is worth considering Arman's aesthetic project in light of Rathje's archaeological one. In collecting and preserving the stuff fellow artist Jim Dine threw away, for instance—and labeling it as such—Arman suggests the capacity of garbage to render a portrait of the person who discarded it. This is precisely the archaeological potential that Rathje mines. As he

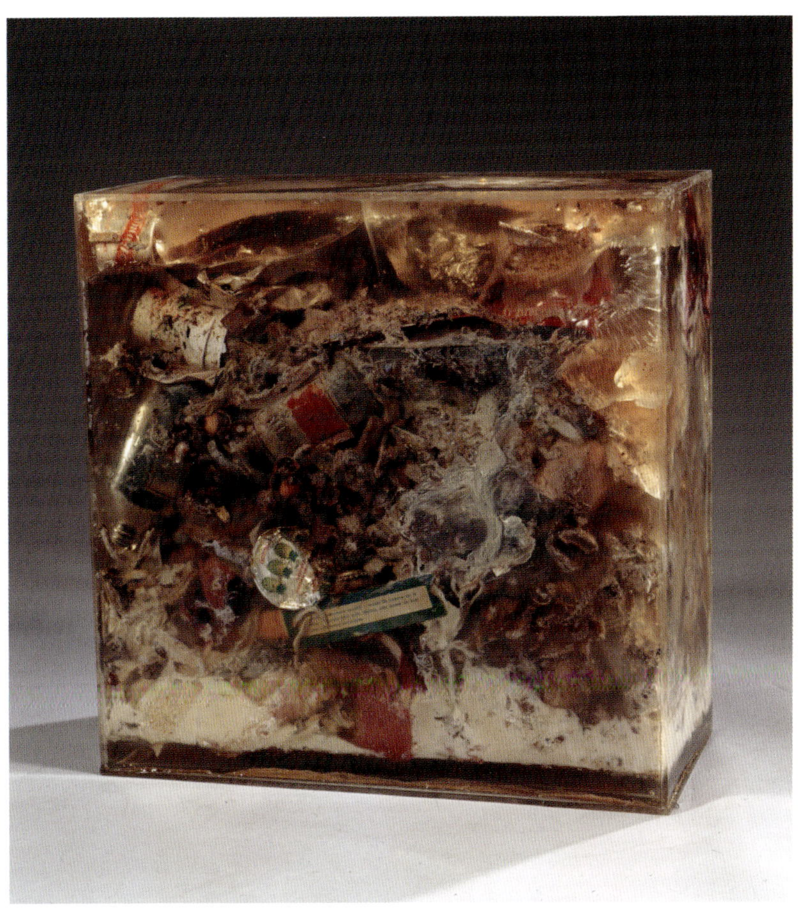

[FIG. 19]
Arman, *Trash Mash No. 1*, 1969. Accumulation of mixed media (refuse) embedded in polyester, 18 × 18 × 10 in. (45.7 × 45.7 × 25.4 cm). Arman Archive Number: 8016.69.010.

puts it in *Rubbish!*, "One of the central tenets of the Garbage Project [is] that what people have owned—and thrown away—can speak more eloquently, informatively, and truthfully about the lives they lead than they themselves ever may." Indeed, the very purpose of artifacts is to render a portrait of a culture, and perhaps as Rathje points out, because this testimony is not filtered through the fears, desires, deceptions, and projections of individual subjects, it has a greater capacity for truth telling than other forms of evidence. In his *Mouse Museum* of 1965–77, sculptor Claes Oldenburg makes a different kind of portrait of a culture by memorializing a collection of kitsch objects, gathered and saved over the years, that seem as though *they would have been garbage* had he not preserved them. The *Mouse Museum* is an intimate freestanding "gallery" whose floor plan is drawn from Oldenburg's stylized rendering of Mickey Mouse [FIG. 20]. Its dark interior is lined with brightly lit inset vitrines holding the collection and two freestanding pedestals where Mickey's eyes would have been in place. Many of the museum's contents consist of novelties in plastic—small figurines, dog toys in the shape of food, anatomical models, and so on. If Arman used plastic to preserve raw garbage as an artifact, Oldenburg interrupts the cycle of overproduction and disposability characteristic of inexpensive plastic items (which are ubiquitous in contemporary consumer culture) by rendering them museum objects. In both cases, the museum serves as the apparatus that can dignify detritus, but conversely there is the urge—as aesthetic as it is archaeological—to *read* rubbish as material culture. What distinguishes this effort (and that of Rathje) from the traditional practice of archaeology is how the materials excavated are contemporaneous with the culture of the artist/archaeologist. The museum's capacity to preserve (as opposed to discard) is matched by its capacity to dignify ordinary, even repulsive objects as valuable forms of cultural property.

In the work of the artist Kelly Jazvac, the metaphoric quality of Arman's conceit of fossilized garbage or Oldenburg's ironic museumification of plastic novelty items gives way to the status of scientific evidence. Along with geologist Patricia Corcoran, and with assistance from oceanographer Charles Moore, Jazvac identified what they call *plastiglomerate*, a conglomerate of stone and sand fused together with plastic on a beach at the southeastern tip of the island of Hawaii. Despite its remote location, Kamilo Beach is a sink for plastic washing up from one of the great Pacific Ocean gyres, and thus was abundantly furnished with polymers that campers or picnickers could unwittingly transform into a new geological phenomenon by lighting bonfires. In a scholarly paper published in 2014, and coauthored by Corcoran, Moore, and Jazvac, the authors suggest that the plastiglomerate might be "a global marker horizon in the Anthropocene." But her scientific contribution is only part of Jazvac's work with plastiglomerates; she has exhibited these "stones" as ready-made sculptures, and they have even been acquired by such institutions as the Yale Peabody Museum in New Haven, Connecticut; Het Nieuwe Instituut in Rotterdam; and Natura Artis Magistra in Amsterdam. In the course of some fifty or sixty years, the dialectic of garbage and artifact has therefore shifted from the realm of speculative fabulation to one of scientific rigor. It is significant that the plastiglomerate now moves between the natural history museum and the institutions of art; this mobility allows it, in its own kind of analytic conglomeration, to fuse speculation with science.

Climates of Disposability

The scale of plastic production since 1950 has exploded—in large part due to its disposability. Single-use plastics are both ubiquitous and easy to overlook due to the modesty of any individual piece of packaging or plastic container that may pass through our hands, and yet a vast proportion of this stuff ends up in landfills, or littering terrestrial or marine environments. In what might be understood as an act of redemption or remediation, several artists have produced environments of plastic components—often creating immersive organic illusions from a material typically associated with waste and toxicity. As in the previous dialectical oppositions between plastic and plasticity that I have considered, such strategies span the period of plastic's expansion, from the mid-twentieth century to the present. Andy Warhol's *Silver Clouds* (1966) invades a room with pillow-shaped, metalized plastic balloons, filled with helium and allowed to float freely among spectators like a fallen polymer sky [FIG. 21]. More recently, as one of Tara Donovan's several sculptures composed of vast numbers of precisely aggregated plastic elements such as straws, mylar tape, or polyester film, her *Untitled* work of 2003 creates a glowing translucent cloud hugging a gallery ceiling, which is entirely composed of Styrofoam cups glued together and illuminated from within. These lyrical atmospheric works literally evoke what art and aesthetics theorist Amanda Boetzkes has called a "climate" composed of plastic. In her 2019 book *Plastic Capitalism: Contemporary Art and the Drive to Waste*, Boetzkes writes, "Plastic produces a climate: it is the medium through which to develop a sensibility within and for an immersive, changing and atmospheric space." In addition to Donovan's work and Jazvac's plastiglomerates, Boetzkes discusses a range of contemporary artists, many from Asia, whose archaeological approach to plastic kitsch or detritus leads to an immersive environment, or a kind of ecosystem. These include the US artist Portia Munson, who accumulates numerous plastic objects of the same hue into delirious accumulations; Chinese artist Song Dong, whose *Waste Not* (2008) set out the contents of his mother's small residence, much of it plastic; and Korean artist Choi

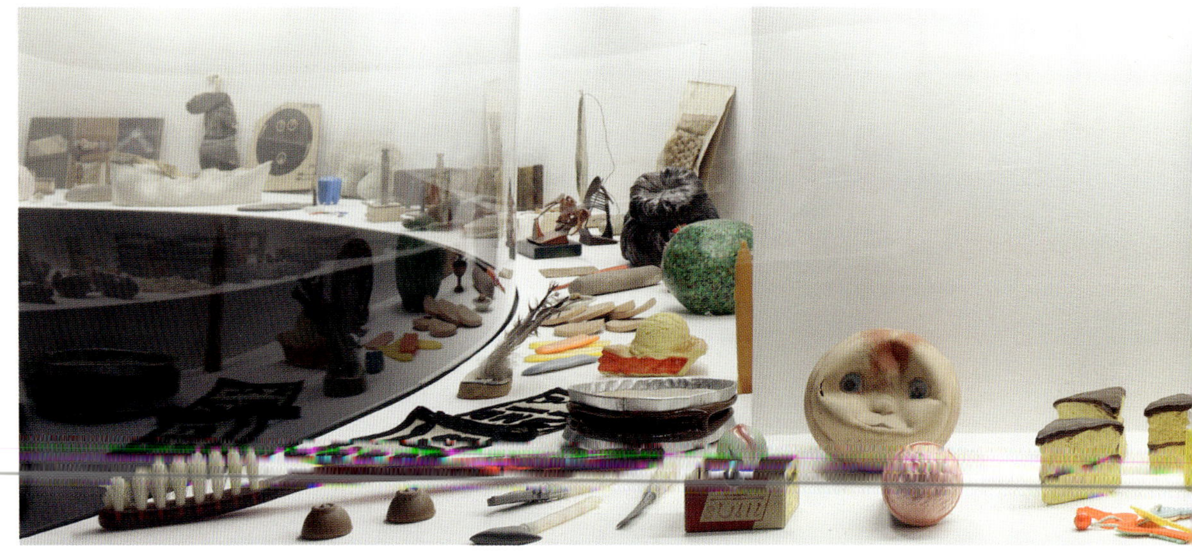

[FIG. 20]
Claes Oldenburg, *Mouse Museum*, 1965–77. Wood and corrugated aluminum, Plexiglas display cases with 385 objects, sound, 103.6 × 378 × 396.4 in. (2.63 × 9.6 × 10.07 m). Installation view. Museum of Modern Art, New York (April 14–August 5, 2013).

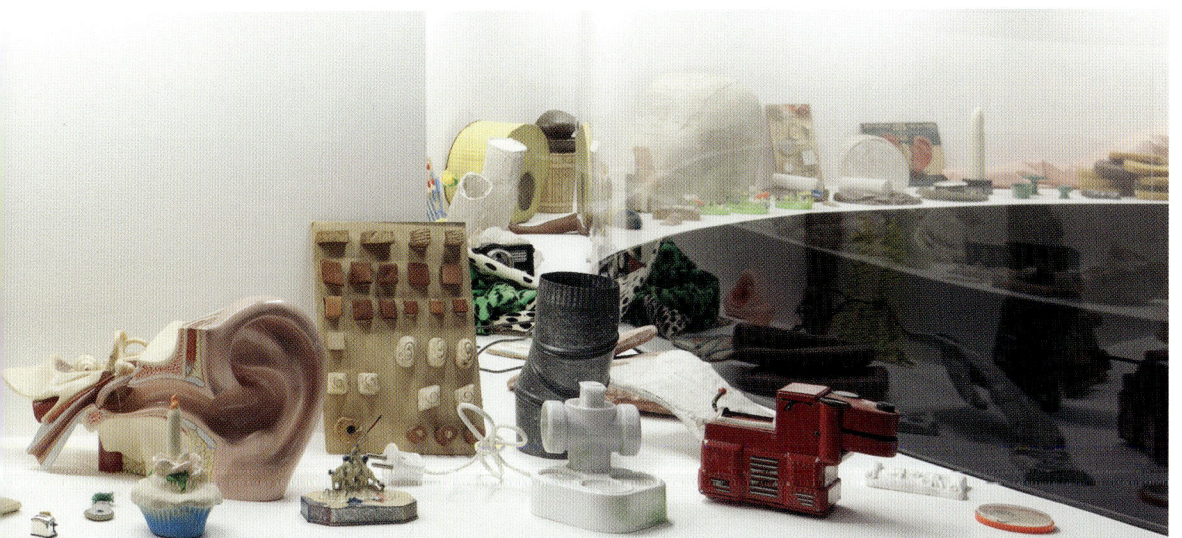

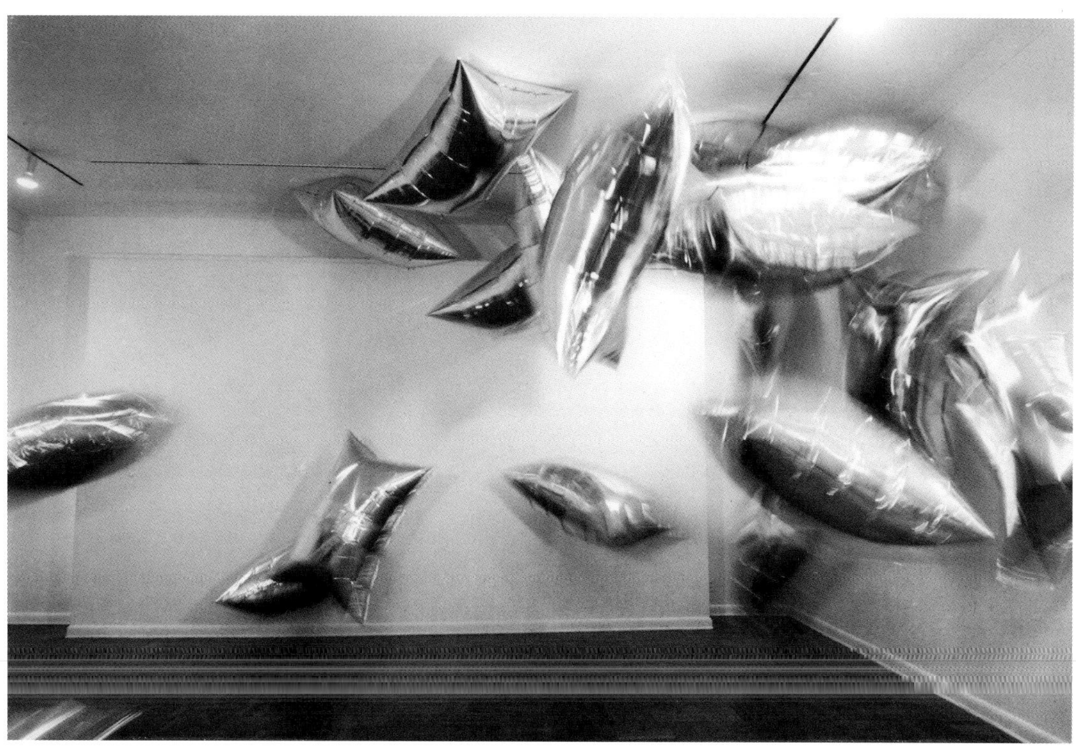

[FIG. 21]
Leo Castelli Gallery, *Andy Warhol: Wallpaper & Clouds*, April 6–7, 1966.

Jeong Hwa, who creates celebratory immersive environments from modest plastic components. While it is undoubtedly a mixed blessing that we live in a "climate" of plastic, it might be more productive to adopt a perspective in sympathy with philosopher and theorist Donna Haraway, who exhorts us to "stay with the trouble" by responding to a changed and distressed environment, not through heroic claims to restore it to some imagined virginal state, but to work *with* the conditions we find ourselves in, toward improving the lives of humans and nonhumans alike. As Liboiron puts it in *Pollution Is Colonialism*, "Purity is not an option here—plastics are already in Land relations." Liboiron's understanding of land relations is drawn from a deep engagement with Indigenous forms of reciprocity between humans and the other life-forms as well as inorganic materials that compose the land. To improve the land relations of plastic, Liboiron first points out that this category includes myriad materials with different production methods, uses, and toxicities. Consequently, they write,

> There are many ways to ensure that other ways of understanding chemicals and pollution flourish. For this reason, plastics are an ideal pollutant to upset dominant norms of pollution—their industrial, intergenerational, and ubiquitous relations make a lot of room for understanding and doing things differently.

Plastic Globe

Globalization introduced plasticity into both procedures of manufacturing and patterns of consumption. The process of making things was stretched and expanded geographically so that the various components that constitute a single product could come from far-flung locations, while global supply chains and internet commerce exploded the presumption that consumption had to take place through a face-to-face transaction. Indeed, most urban consumers have little idea where the products they use come from or where they go once those things have been discarded as waste (after shorter or longer productive lives). Thomas Princen, a theorist of ecological sustainability, has described this condition as *distancing*. In a 2002 essay titled "Distancing: Consumption and the Severing of Feedback," he observes, "'Distance' is the separation between primary resource extraction decisions occurring along four dimensions—geography, culture, bargaining power, and agency." But the consequences of such distancing are significant, as Princen elaborates: "Severed ecological feedback and diminished accountability via the distance of multiple agency also separates *rights* for resource use from *responsibilities* of that use." What he means by "severed ecological feedback" is that the environmental devastation of say, oil extraction, is so remote from the vast majority of those who consume oil (or the plastic that some 6 to 10 percent of oil is transformed into) that they tend not to respond with outrage or political resistance to the deleterious effects of that extraction in the course of their everyday lives. In other words, the right to consume oil (or plastic) is detached from the responsibility to ensure that its extraction and use are not environmentally disastrous. Responsibility is submitted to a kind of plasticity that benefits corporate interests and veils ecological harm. Political economist Jennifer Clapp has made a complementary argument regarding the distance between what we discard and where it ends up. Indeed, the rise of globalization is roughly contemporaneous with the mid-twentieth-century explosion of waste, but privileged consumers of richer nations seldom directly confront the mass of stuff they throw out. As academic Kate O'Neill has argued, due to the plasticity enabled by global markets, waste itself is now traded so that the refuse of the affluent tends to flow into poorer and disadvantaged communities both within a single nation, and between rich and poor countries. Like Princen, O'Neill speaks of distancing. In her 2019 book *Waste*, she contends, "The negative impacts of wastes have always disproportionately affected the most economically disadvantaged, who often belong to racial or ethnic minority groups, but geographic distancing of waste disposal from the point of production has exacerbated these risks." The plasticity of distancing is how rights are claimed without assuming responsibility.

In the realm of art, the plasticity of distance and proximity performed by globalization has met with multiple responses. One of these centers on activism that seeks to heighten the contradiction of oil companies' patronage of the arts. In 2010, for instance, the activist group Liberate Tate began to protest BP's sponsorship of the four Tate galleries including Tate Britain and Tate Modern. Art patronage is pursued by large corporations such as BP as a means of burnishing—or some might say laundering—their reputations by creating a greater conceptual distance between economic pursuits that cause real ecological harm and the aspirational projection of a benign corporate identity. The dramatic actions devised by Liberate Tate sought to collapse this distance—and succeeded in doing so to the extent that Tate and BP ended their relationship in 2017 after twenty-six years of a partnership. In the 2010 performance *License to Spill*, for example, a group of uninvited guests (members of Liberate Tate) attended the annual summer party at Tate Britain, which was dedicated to celebrating twenty years of BP sponsorship. Under their floral party dresses, these initially unobtrusive intruders carried sacks of molasses, which they released onto the museum's floor in a not-so-abstract simulation of BP's massive oil spill in the Gulf of Mexico the previous April. Shortly after this initial action, twelve additional Liberate Tate performers, veiled in black, dumped more of the molasses from

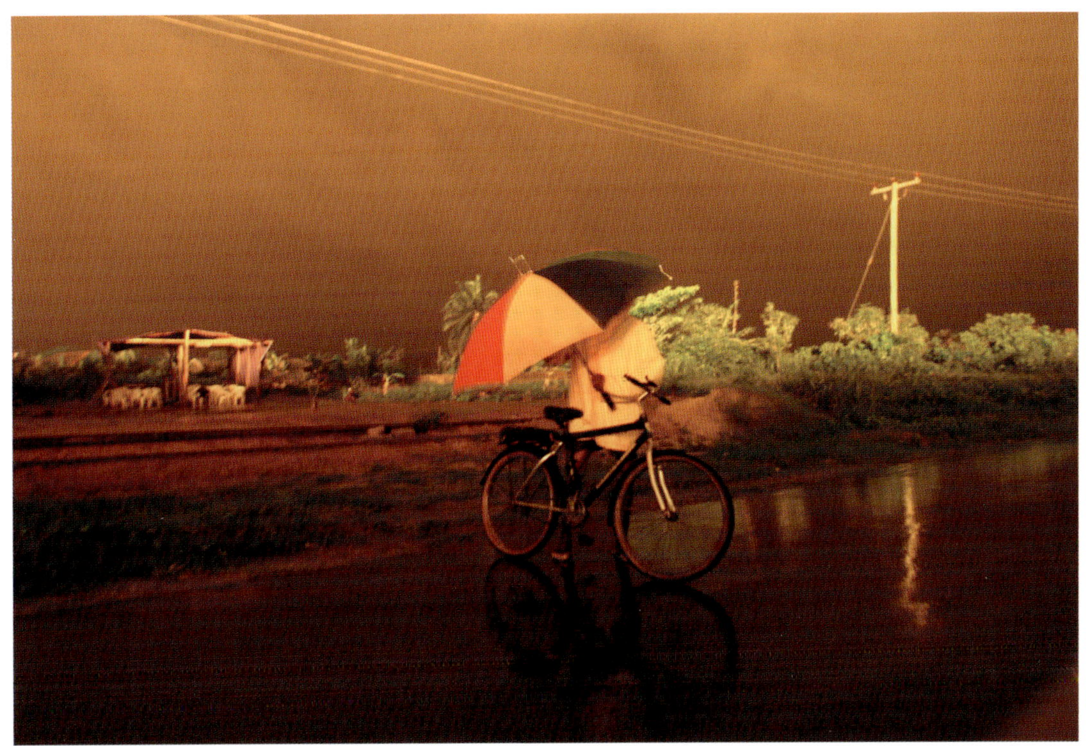

[FIG. 22]
George Osodi, *Oil Rich Niger Delta*, 2003–2007.

containers emblazoned with BP's "helios" logo. The strategy here was to bridge the ideological divide between corporate philanthropy and industrial disaster by creating an oil spill in the British art world to parallel the one across the Atlantic in the Gulf of Mexico. This mode of plasticity functioned as a deeply material return of the repressed, which could bend one geographic location into another.

Individual artists have developed other aesthetic means of countering the distancing effects that sever rights from responsibilities. George Osodi's photographs of the Niger Delta, for instance, document the collision of two worlds in a region of Nigeria where foreign corporations extract resources while impoverishing local communities [FIG. 22]. The plasticity that Osodi represents is ecological; he documents the emergence of a new ecosystem resulting from the transformation of the third-largest wetlands in the world into a dystopian expanse of oil fields. Osodi's series of over two hundred photographs pictures this encounter between Western oil extraction and African agriculture through various kinds of juxtaposition, including the image of storage tanks from a Mobil oil "farm" looming over a group of boys playing soccer, photographs of villagers siphoning fuel from overturned tankers, or the portrait of a woman gazing at a leaking oil well on the way to her farm. But for me the most haunting images of Osodi's essay are those that capture "orange light"—the unearthly glow of gas flares that tint the Niger Delta's atmosphere by day and night. Open flares are used to burn off the excess gas released in oil production; they not only waste a valuable resource but lead to significant environmental harm too. As Nigerian photographer Tam Fiofori puts it in a monograph dedicated to Osodi's work, "The flares cause acid rain that in turn poisons both the soil and the creeks stunting the growth of agricultural produce and killing most marine life. . . . The gases burn the skin of humans and cause skin cancer. They also effect the eyes and digestive system." Some of the most powerful works in Osodi's series register the nighttime effects of the gas flares—an orange-hued world that is both beautiful and terrible. These are images of a toxic plasticity, a new weather made from the transformation of wetlands into a fiery hellscape.

The toxic plasticity of oil that Osodi's photographic essay documents is complemented by practices of reparative plasticity in what has been called "plasto-art" by Nigerian artist Ifeoma Anyaeji. As academic Jennifer Wagner-Lawlor wrote in 2018 essay titled "Poor Theory and the Plastic Pollution in Nigeria: Relational Aesthetics, Human Ecology, and 'Good Housekeeping,'" "There is a 'second-order' crude reality [to the problem of oil extraction]: the 'oil spill' called plastic pollution. . . . But the gold-rush for plastics companies in Nigeria has just begun." Anyaeji harvests plastic bags and bottles from this "'oil spill' called plastic pollution,"

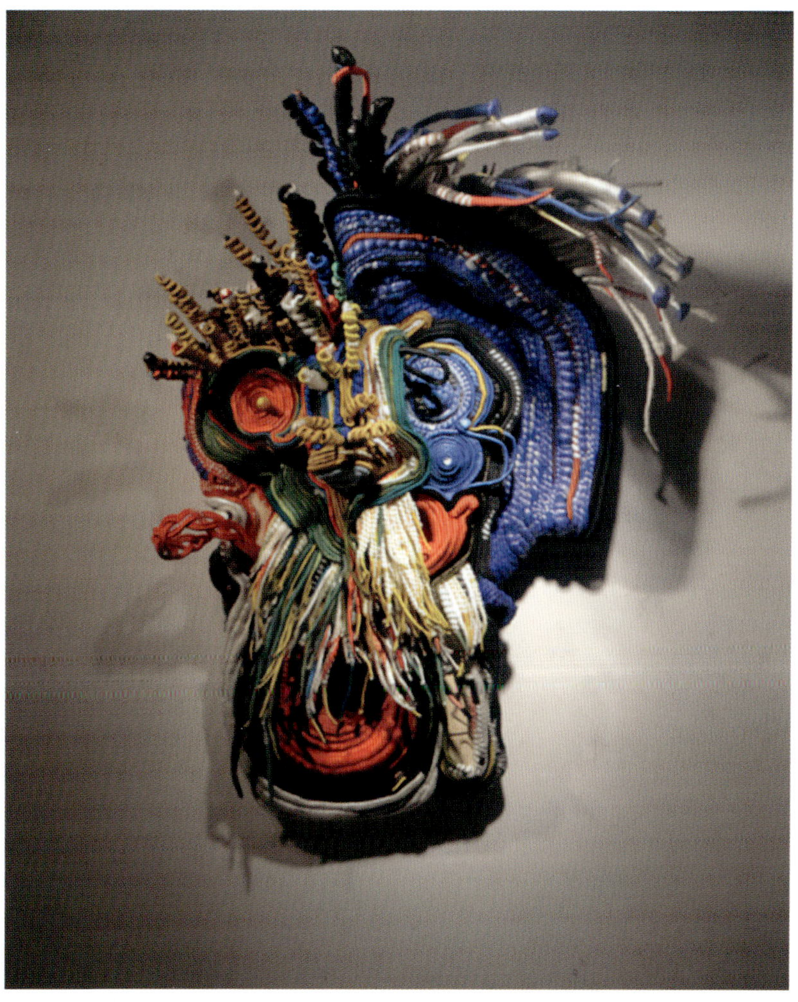

[FIG. 23]
Ifeoma Anyaeji, *Ghangha Ọ ha mmili* (*Ghangha the rain-maker*) (front view), 2019. 50.8 × 33.5 × 63.1 in. (129 × 86 × 160.2 cm).

transforming some of her gleaned materials into "plasto-yarns" that she weaves into works through formal procedures inspired by Nigerian forms of hair plaiting or threading [FIG. 23]. By applying traditional techniques of adornment to urban waste, she initiates a plasticity opposite to that pictured by Osodi: instead of showing how traditional forms of life are assaulted by corporate oil extraction, the traditional form of hair plaiting absorbs and transforms contemporary plastic waste. Anyaeji uses her plasto-yarn to build powerful abstract sculptural reliefs that frequently evoke monumental masklike objects and sometimes expand into home furnishings. As the artist eloquently puts it in her online "Artist's Statement," her plasto-art is not meant to simulate tradition but instead to activate it within contemporary global conditions: "Its intent is to not aid the growing obliteration of traditional African aesthetic values by global cultural pressures. Rather, to encourage the sustenance of these values within global cultures. Plasto-art is an exploration on the intensity and adverse magnitude of contemporary value systems and my interpretation of traditional material culture reuse in Nigeria."

The Plasticity of Plastic

In this chapter, I have approached the aesthetics of plastic in terms of a material dialectic—one in which, following Malabou, matter gives, assumes, and destroys form. What this exploration of plastic's plasticity enables is not only a speculative art history of a particular species of stuff—plastic—but also an alternate framework for deriving meaning from art's materiality. The modern concept of the medium regulates matter by furnishing it with a Kantian set of a priori rules; there is a perpetual analytic reduction of what, for instance, a painting can be. In short, the openness of the dialectic is foreclosed as a teleology. But the dialectical behavior of plasticity, according to Malabou, is *explosive* (she speaks of plastic explosives). Rather than tending toward a preordained outcome, it may engender startling transformations and raise aporetic paradoxes. This perspective allows us to apprehend a deeply material kind of formalism in which an artistic process absorbs and works through the social lives of its constituent material. An account of plastic in art is thus inextricable from the paradoxical postwar history of plastic as at once a source of toxicity and potential tool for reducing risk as well as enhancing medical technologies and food preservation. I have attempted to show that charting what interdisciplinary scholar Heather Davis has called the "plasticity of plastic" allows us to reduce the distance between its social antinomies—by bringing an oil spill into a gallery or weaving plastic waste found on contemporary Nigerian streets into a sculpture using traditional techniques of hair plaiting. From this perspective, a material is not just stuff but instead a bundle of social relations inextricable from the "behavior" of that material both within art and beyond. A material dialectic carries the capacity to explode these social relations but also the potential to repair them. I say potential because it is exactly that: a dialectic is not a teleology; its outcome cannot be known in advance.

Case Studies in Plastics

The Conservation Treatment of a Naum Gabo

Elena Torok

In the early twentieth century, the commercialization of plastics transformed industry, especially the art world. These new and innovative materials had properties that rendered them ideal for sculpture; they were lightweight and inexpensive, sold in a range of different colors and opacities, and could be easily cut, shaped, and molded. Although artists began readily incorporating plastics into their work almost immediately, it was not understood until many years later that some of these materials had inherent chemical instabilities that caused them to age poorly over time.

From the start of his career, artist Naum Gabo (discussed in the introduction) used plastics in innovative ways. Some of his earliest sculptures, constructed while he was living in Russia (1917–22) and Berlin (1922–32), were made from cellulose esters, semisynthetic polymers derived from chemically treated cellulose. These materials, including both cellulose acetate and cellulose nitrate, are now considered to be among the most unstable plastics found in museum collections today. In addition to issues with chemical instability, Gabo's early plastic sculptures are vulnerable to damage because they are made from thin and lightweight sheets of plastic, and have fragile plastic-to-plastic joins. As the artist moved many times during his life and exhibited his works in exhibitions all over the world, it was not uncommon for breaks to occur in transit. Gabo witnessed the deterioration of his early plastic works during his lifetime, causing stress to both him and his collectors. Some of the artist's early plastic sculptures no longer survive, and of those that do, many are now in significantly altered condition.

For decades, scientists and conservators have worked to understand the aging mechanisms of early plastics, and this collective research has allowed for multiple advancements in their conservation and care. Although it has been determined that the chemical breakdown of cellulose esters can be slowed with strict control of surrounding environmental conditions, it cannot be fully stopped. Therefore, identifying and understanding these unstable plastics in works of art is critical in planning for their long-term preservation.

The recent conservation treatment of Gabo's Ivory Rhodoid *Constructed Head No. 2* [FIG. 24] at the Dallas Museum of Art (DMA) involved new research into the sculpture's history, plastic composition, and previous repair. This research provided insight into why this off-white plastic sculpture, now a century old, may survive in more stable condition than many of Gabo's transparent plastic sculptures from the same era. Research also informed decisions made in treatment, which improved condition issues and allowed this important work from the artist's early career to return to public display after multiple years in storage.

Gabo's Early Career

Gabo was born Naum (Neemia) Borisovich Pevsner in Briansk, Russia, in 1890.

He grew up with strong foundations in the sciences, which were established during his youth in Russia (1890–1910) and education in Germany (1910–14). Gabo began making art by 1915, while he was in his mid-twenties and living in Norway with his brother, artist Antoine Pevsner (1884–1964). At this time, he also started using the surname Gabo as a way to differentiate himself from his brother. Gabo's deep-rooted interests in science, technology, engineering, and materials served as major influences in his artwork throughout his life and career. Some of Gabo's earliest sculptures were figurative constructions of torsos and heads, built in a variety of materials using his stereometric method, which relied on the use of space to build form. These sculptures included *Constructed Head No. 1* (first modeled in cardboard and then built in plywood in 1915), *Constructed Head No. 2* (first modeled in cardboard and then constructed in galvanized iron painted with yellow ochre paint in 1916), and *Constructed Torso* (first modeled in cardboard and then built in sheet metal covered in sand in 1917–18).

In 1917, after the outbreak of the Russian Revolution, Gabo and Pevsner moved back to Russia, where both continued making art. In 1920, Gabo published the *Realistic Manifesto*, a seminal text in the art movement of constructivism that was cosigned by Pevsner.

During these years, Gabo began using plastics in his work and constructing forms that were more abstract than figural. Some of these early plastic works included *Constructed Head No. 3* (*Head of a Woman*) in 1917–20, *Model for Column* in 1920–21, *Construction en Creux* in 1920–21, and *Construction in Space C* in 1919–21. The first two sculptures survive intact but in fragile condition, while the latter two are now lost.

[FIG. 24]
Naum Gabo, *Constructed Head No. 2*, 1923–24. Ivory Rhodoid, 16.5 × 16.75 × 12 in. (41.9 × 42.5 × 30.5 cm). Dallas Museum of Art, Texas. Edward S. Marcus Memorial Fund, 1981.35.

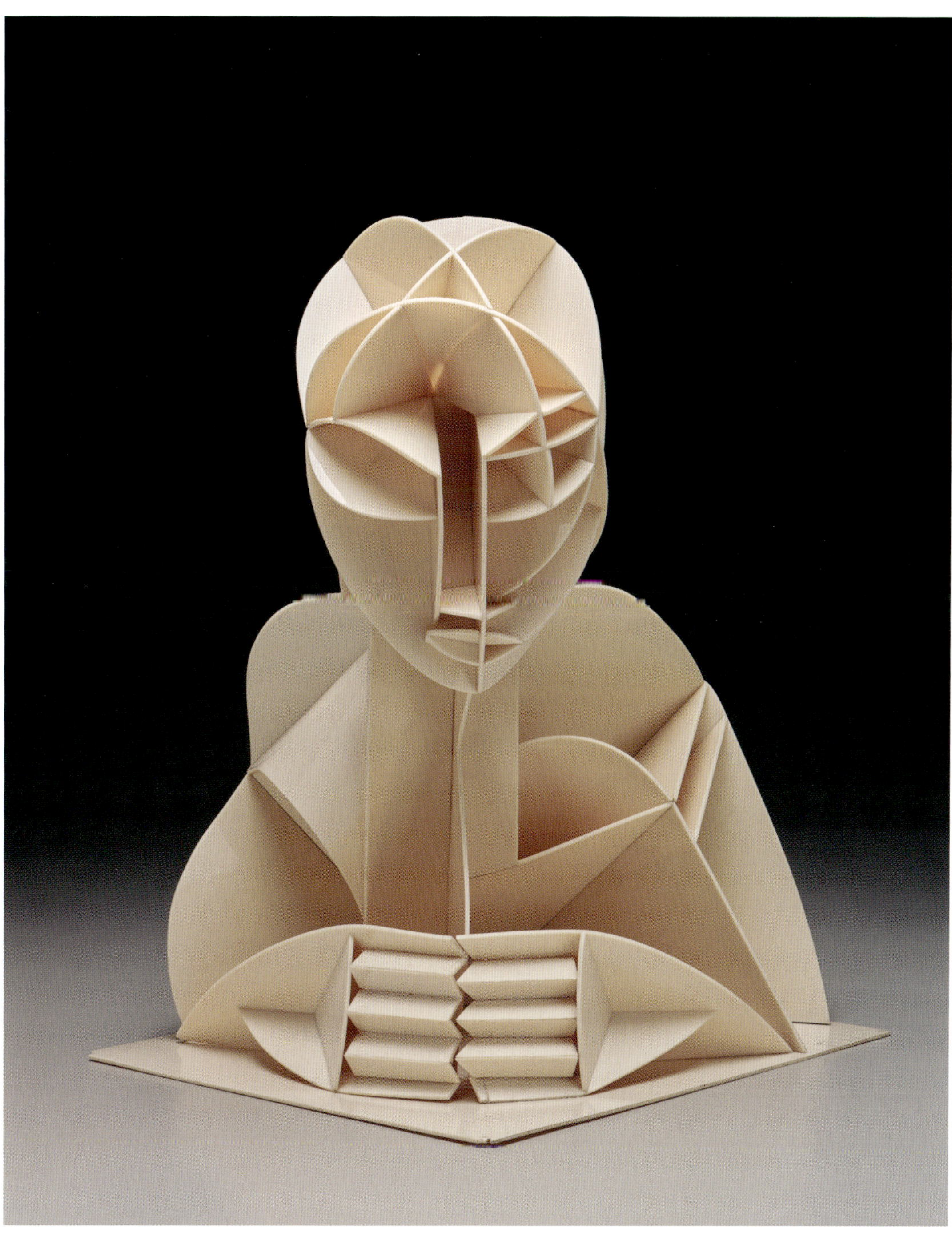

Gabo in Berlin and the Ivory Rhodoid *Constructed Head No. 2*

In 1922, Gabo left Russia for Germany to help install the *First Russian Art Exhibition*, organized by the Soviet government at the Galerie van Diemen in Berlin. The exhibition included at least eight of Gabo's sculptures, including the galvanized iron *Constructed Head No. 2* that he made in 1916. The sculpture went missing, however, shortly after the close of the exhibition and was not recovered until the early 1960s.

Gabo must have still had access to a set of templates, as he made another version of *Constructed Head No. 2* in 1923–24, but instead of using metal, this new version was built from white plastic. This version is now in the DMA's collection. In the 1950s, the artist noted that the plastic used for *Constructed Head No. 2* is Ivory Rhodoid, which he purchased in sheets while living in Germany.[1] Its initial exhibition date has always been difficult to ascertain. It was exhibited in Vienna in late 1924 but was also likely included in the *Constructivistes Russes: Gabo et Pevsner* exhibition at the Galerie Percier in Paris earlier that year. Gabo owned the sculpture until his death in 1977. He left Berlin for Paris in 1932 and then later moved to England in 1939 after the outbreak of World War II. In 1946, he moved again from England to the United States and was granted citizenship there in 1952. The Ivory Rhodoid *Constructed Head No. 2* remained in Gabo's possession throughout all of these moves. After his death, it was sold to the DMA by the artist's family in 1981. In his later career, Gabo revisited the *Constructed Head No. 2* form frequently. Between the early 1950s and 1975, the artist built five additional versions of this sculpture, all but one in larger dimensions, and all in metal. The DMA sculpture is the only plastic version of *Constructed Head No. 2* that is known to exist.

The Making of *Constructed Head No. 2* in Ivory Rhodoid

Gabo's identification of the plastic as Ivory Rhodoid provides insight into its potential material composition. Rhodoid, a trade name for cellulose acetate, was first manufactured by the French company Société des usines chimiques du Rhône in 1917 and was sold commercially in Europe in a range of different colors. The off-white sheets Gabo used to make *Constructed Head No. 2* imitate natural ivory because they contain faux Schreger lines, faint striations of alternating off-white tones that run throughout the substrate, and mimic the naturally occurring surface features found in real elephant and mammoth ivories.

Gabo was using multiple different commercial plastics in his sculptures at this time and frequently used the word *rhodoid* as a descriptive term for plastics in general, no matter their chemical composition. As of 2023, scientific techniques had not confirmed that the material used to construct the DMA *Head* was cellulose acetate. The sculpture's appearance, condition, and odor, though, are consistent with an aged cellulose ester.

Studies of Gabo's plastic sculptures from the 1920s have determined that he was using plastic sheets in his artworks that have since been identified as either cellulose acetate or cellulose nitrate.

Despite the age of the DMA *Head*, it is still in stable condition, especially as compared to many of Gabo's other early plastic works. For example, Gabo's *Construction in Space: Two Cones* (1927), constructed largely from sheets of clear, red, and black cellulose acetate, rapidly deteriorated in 1960 when the vitrine was opened at the Philadelphia Museum of Art (PMA). *Model for Column* (1920–21), constructed from sheets of transparent cellulose nitrate, is also now severely yellowed. The DMA *Head* exhibits some yellowing and discoloration (particularly in enclosed areas of the form, where off-gassed acidic pollutants have had significantly less airflow) along with some warping and planar deformation (in the sculpture's lower proper left shoulder), but despite this, the sculpture is structurally sound and intact overall.

The plastic's good condition may be related to a combination of chemical and environmental factors. In 2017, multiple locations were examined by the DMA's conservation department using a handheld X-ray fluorescence spectrometer (an instrument that can nondestructively identify elements without the need for physical sampling) to learn more information about the materials embedded in the plastic matrix. All examined areas were found to contain both zinc and titanium. Although additional scientific analysis would be necessary to determine the specific compounds associated with these elements, they may be related to the sculpture's color, as zinc oxide and titanium dioxide were both used commercially as well as industrially as white colorants and opacifying agents during the twentieth century.

In scientific studies conducted at other museum collections, plastic objects comprised of white, off-white, or faux ivory cellulose nitrates have been found to degrade more slowly than cellulose nitrates that are either uncolored or not white. For example, MoMA's off-white plastic *Constructed Head No. 3* (*Head of a Woman*) is in a similarly intact condition as the DMA *Head*. In 1991, it was determined to consist of cellulose nitrate with zinc oxide as a principal opacifying pigment.[2] In 2007, the scientific analysis of *Construction in Space: Two Cones* at the PMA determined that a white polygon, the most intact element of an otherwise severely degraded sculpture, was also made from cellulose nitrate with zinc oxide as an opacifying pigment.[3] It has been theorized that zinc oxide may slow the breakdown of cellulose nitrate by neutralizing acids formed in the polymer matrix. Though these published studies have been specific to cellulose nitrate, it is possible that zinc oxide may have a similar neutralizing effect on the breakdown mechanisms of other cellulose esters (like cellulose acetate) as well. Gabo's choice of color for his plastic *Constructed*

Head No. 2 seems to have significantly influenced its intact condition today.

Fabrication

The DMA's *Constructed Head No. 2* appears to have been fabricated in a manner consistent with Gabo's documented working methods at the time. It is comprised of over thirty individual cut plastic elements that have been assembled to make the overall form, all a uniform thickness of 2.25 millimeters. Pencil marks are visible around many edges, which appear to remain from Gabo's tracing of each element from a set of templates on the larger plastic sheet. After tracing, Gabo cut each shape to size. The material was too thick to cut with scissors, so Gabo may have used a handsaw, such as a coping saw or hacksaw. Straight lines may have been scored with a blade and snapped too. Many edges are smoothed or finished, and some are also beveled. The thermoplastic qualities of this material would have allowed Gabo to form curved elements using heat, perhaps in combination with a mold to form curvatures of a specific shape. No glue is present in the original joins, indicating that Gabo joined plastic elements together using solvent. Gabo often used solvent to create joins in his early plastic works, but began incorporating adhesives later on to reduce his exposure to toxic chemicals. Evidence of solvent joining is visible in an area where a vertical plastic element is now missing from the sculpture's proper right eye; the surface appears chemically disrupted along the join line, and light scoring marks are visible as well.

Condition Issues

By 2017, the plastic substrate was still in generally stable condition, especially as compared to many of Gabo's other early plastic works. Yet the sculpture had been off view for multiple years because the adhesive along most of the joins was discolored and aged, which altered the sculpture's overall appearance and translucency, and compromised its structural integrity in some locations. Since Gabo formed plastic-to plastic joins in his earliest works using solvent, all of the glue was believed to have been applied later, perhaps during restoration campaigns. As Gabo repaired and restored many of his own works during his lifetime, research was necessary to understand when these repairs may have been applied and by whom.

At least two distinct adhesive types were present: a soft, dark yellow adhesive applied to most joins overall, and a harder, shinier, clear adhesive located only in the figure's hands. Three small white plastic fills were also present: two in the figure's proper right shoulder (located where two different plastic elements met the base), and one at the top of the figure's head. DMA object files did not indicate the sculpture had received any major structural conservation treatment since it entered the collection, so the adhesive appeared to have all been applied prior to the sculpture's acquisition in 1981.

Through archival research and conversation with Gabo's long-term studio assistant, Charles Wilson, it was determined that the sculpture had been repaired at least three times: once by Pevsner in 1952, again by Gabo later that same year, and then by Wilson in 1981 (a few years after the artist's death). The 1952 damage occurred while the sculpture was in transit to the *L'Oeuvre du XXe siècle* exhibition at the Musée National d'Art Moderne in Paris. James Johnson Sweeney, curator of this exhibition, telegrammed Gabo as soon as the damage was noted to inform him and ask for advice on repair so it could still be exhibited.[4] Despite careful packing by Gabo himself, the head arrived separated from the base, and unspecified damage was present in the hands.[5] Gabo received the telegram four days later and replied to Sweeney right away:

> I cannot understand how the damage could happen. The only man who could repair it is Antoine Pevsner. I am writing to him at once authorizing him to do the repairs. . . . I don't know anybody else in Paris but if for some reason, Antoine is not there, I think that if the damage is slight, a good workman may temporarily fix the head onto the base with acetone. Temporary fixing is done very easily with a small water-colour brush and a bit of acetone as seccetive [*sic*]. Press the parts together, not too hard, and hold for a few minutes.[6]

By the time Sweeney received the message, Pevsner had already become aware of the damage to Gabo's *Head* while checking on the condition of one of his own sculptures in the same exhibition. Without waiting for a response or instruction from Gabo, Pevsner moved forward with the repair. Pevsner's wife, Virginie, reported to Gabo that Pevsner did not repair the sculpture with acetone but rather with a strong glue.[7]

Pevsner's work allowed the sculpture to be shown in Paris (and also in the exhibition's second venue, Tate Gallery in London), but his repairs were only ever intended to be temporary, so Gabo would still need to perform additional work on the sculpture when it was returned to his studio in Connecticut later that year.[8] In fall 1952, Gabo wrote to both Sweeney and Walter Lenox (of insurance provider Toplis & Harding) to note that the base and lower-right part of the figure's shoulder both needed replacement, but he was having difficulty sourcing Ivory Rhodoid in the United States.[9] It can be presumed that Gabo was eventually able to find some, as today the base is intact and matches all other plastic elements in both thickness and color. He may have also added the two plastic fills along the bottom edges of elements comprising the figure's right shoulder at this time. Gabo's signature is located in Cyrillic on the base's proper left side [FIG. 25]. The artist did not start regularly signing his work until the later part of his career. If the base was replaced in 1952, the signature must have been added sometime after this date.

In fall 1953, at the *Alexander Calder, Mobiles—Naum Gabo, Kinetic Construction and Constructions in Space* exhibition at the Wadsworth Atheneum Museum of Art, Gabo's sculpture was exhibited for the last time during his lifetime. An archival photograph from the exhibition contained in the museum's archives confirms that the sculpture was structurally sound and once again in exhibitable condition, but a vertical plastic element is missing from the figure's right eye. It is unclear if this element was lost during the 1952 damage, but a photograph of the sculpture published in the *Washington Daily News* on October 9, 1948, shows the right eye as intact, so the loss must have occurred sometime between 1948 and 1953. It is not known if Gabo noted this damage or ever attempted to repair this area, although the element was still missing when the work was acquired by the DMA in 1981.

It is not clear why Gabo stopped exhibiting the Ivory Rhodoid *Head* after 1953. In the mid-1950s, however, it is known that he traced new templates from the sculpture to make a new version of the *Head* in a phosphor bronze. According to Wilson, Gabo was never fully satisfied with this version. Since the Ivory Rhodoid *Head* was not disassembled to take tracings, the dimensions of the phosphor bronze version were much less accurate than other versions. Notably, the phosphor bronze version has the vertical element in the proper right eye that the Ivory Rhodoid version is missing.

By 1965, the Ivory Rhodoid *Head* was moved to long-term storage at Cohen & Powell in New Haven, Connecticut. After the artist's death, it was moved again to a different storage facility in London, where it was first viewed by Dr. Steven Nash, who was appointed the assistant director and chief curator of what was at the time called the Dallas Museum of Fine Arts (now the DMA) in 1980.[10] In 1981, the museum acquired the sculpture, and Nash worked with Gabo's daughter, Nina, to send it to Wilson for cleaning and repair prior to acquisition. At this same time, Wilson was already actively working closely with Gabo's family to repair and restore a number of the artist's sculptures, particularly a group that had been rediscovered in pieces in Gabo's Connecticut studio after his death.

In 2017, Wilson recalled that when the sculpture arrived in Gabo's studio in 1981, adhesive in many of the joins was gelatinous and exhibited some kind of adverse interaction with the plastic. This old adhesive, presumably applied by Gabo, had even caused the softening of some of the plastic edges over time. Wilson took the sculpture apart, removed the

[FIG. 25]
Conservation images of Naum Gabo's *Constructed Head No. 2.* (1923–24), 2020. Dallas Museum of Art, Texas. The sculpture's proper left side, before (top) and after (bottom) conservation treatment to stabilize joins and reduce aged adhesive from a previous repair. Gabo's signature in Cyrillic is located on the front of the sculpture's base.

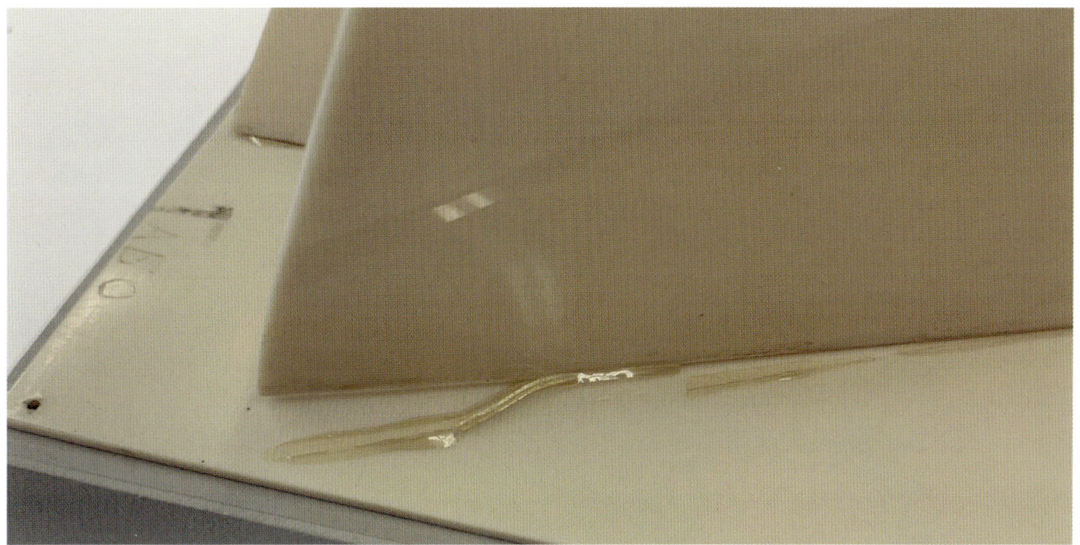
GABO

old adhesive, and trimmed any damaged areas of plastic. He let the pieces sit for two or three days before reassembling the sculpture with glue. Wilson also remembered that a small piece of Ivory Rhodoid was missing from the face; he couldn't recall exactly where this loss was located but was able to find old Ivory Rhodoid stock in Gabo's studio that he used to make the repair.[11] Had it not been for Wilson's intervention and repair after Gabo's death, the sculpture might not have survived to the present, especially not in this same intact condition. Although no photos exist, Wilson's recollections were critical in understanding that much of the adhesive that was present overall in 2017 had been applied in 1981.

In 1985–86, the Ivory Rhodoid *Head* was exhibited in *Naum Gabo: Sixty Years of Constructivism*, a major retrospective organized by Nash; the exhibit traveled to six venues in Europe and the United States as well. Although the sculpture has been on display at the DMA many times in years since, this was the last time it ever left the museum on loan. The sculpture no longer travels because of its age, material composition, and fragile condition. In order to preserve Gabo's surviving early plastic works for as long as possible, museums around the world have been forced to make similar decisions.

Conservation Treatment

The vast information gathered on the sculpture's material composition and repair history helped inform conservation treatment. Research determined that there were at least three previous major repair campaigns: one by Pevsner in 1952, one by Gabo in 1952, and one by Wilson in 1981. Since all repairs had historic importance, any decisions to remove or reduce them needed to be carefully considered, in tandem with an assessment of the sculpture's structural and aesthetic needs.

Since Wilson's 1981 adhesive was still effectively holding the sculpture together yet also darkly yellowed and discolored, it was mechanically reduced as much as possible while still leaving all joins intact. Where additional structural support was needed, joins were reinforced with a clear, reversible adhesive that has long-term aging properties. The clear adhesive in the figure's hands (presumably applied before Wilson's intervention, and perhaps by either Pevsner or Gabo in the 1950s) was still in stable condition so it was left in place. Moreover, although the exact dimensions of the missing vertical piece in the figure's right eye were known, it was eventually decided not to attempt replacement. Any material considered for this repair would need to match the color and translucency of the surroundings exactly, and have no risk of chemically interacting with the adjacent aged Ivory Rhodoid. As the loss was not overtly noticeable, and Gabo was known to have exhibited the sculpture even after this loss occurred, the area was ultimately left as is.

Planning for Long-Term Preservation

Before *Constructed Head No. 2* went on view again after conservation treatment, there was much internal discussion among conservation, curatorial, collections, and exhibitions departments to plan for both safe display and long-term preservation. Although the plastic is still intact, it is still slowly and irreversibly deteriorating. Therefore, long-term environmental planning for storage and display was considered in order to extend the sculpture's life for as long as possible. All cellulose esters are known to deteriorate in time, but the rate of deterioration can be slowed considerably by controlling an object's surrounding environment. Other museum publications and case studies serve as helpful guidelines on the storage as well as display of Gabo's early works.

Light

Long-term exposure to light, particularly in the UV range, can cause the yellowing and discoloration of cellulose esters over time. Light exposure is likely largely responsible for the current yellowed appearance of the sculpture's Ivory Rhodoid substrate. While on display, gallery lighting is filtered from all UV light and normal light levels are kept as low as possible. Since light damage is cumulative, a decision was made to periodically rotate *Constructed Head No. 2* from permanent view so that it can rest for stretches of time in dark storage.

Relative Humidity and Temperature

The deterioration of cellulose esters occurs more rapidly under conditions of higher relative humidity and temperature levels, and damage can be exacerbated by rapid relative humidity or temperature swings. Periods of high or fluctuating heat or relative humidity may have contributed to the warping of the sculpture's shoulder over time. Environmental relative humidity and temperature levels in both the galleries and storage at the DMA are kept at stable conditions as well as within normal museum best practice standards. While the sculpture is off view, the DMA does not have access to cool storage, but the piece is housed in an area that is as cool as possible.

Pollutants

Enclosing early cellulose esters that are actively off-gassing like this one can be problematic, as certain off-gassing compounds will contribute to the degradation of the plastic itself. For display, given the fragile nature of the sculpture, its light weight, and its relatively small size, it did not seem possible to offer visitors the opportunity for close looking and still provide sufficient protection from touching or handling without the use of a case. Therefore a specialized case was fabricated with vents in the Plexiglas bonnet that would allow for passive air exchange while the object was on display. Additional air exchanges were facilitated by the conservation department during nonvisitor hours, either by use of a HEPA-filtered vacuum or the temporary removal of the bonnet. When the sculpture is off view, it is stored on an open shelf and can be partly covered in a way that protects it from dust while still allowing for passive airflow.

Gabo on Fragility

By the time Gabo was living in the United States in the 1950s, the impermanence of his materials and fragility of his sculptures had been noted by museums, scholars, and critics. His fears on how the perception of fragility would have a negative impact on his legacy have been documented in archival correspondence.

In 1951, Gabo was furious when Andrew Carnduff Ritchie (then director of painting and sculpture at MoMA) did not include him in his *Abstract Painting and Sculpture in America* exhibition, particularly because Ritchie also criticized Gabo's work in the associated publication, *Sculpture of the Twentieth Century* (1952). Ritchie wrote that in contrast to Pevsner, Gabo "prefers translucent materials like glass or plastic combined at times with a harp-like gut stringing. What he gains thereby in lightness and airiness he loses, alas, in permanence and stability." Ritchie later became director of the Yale University Art Gallery in 1957. In 1965, Gabo was again furious when the Yale University Art Gallery chose to decline the loan of *Construction in Space, Balance on Two Points* to multiple venues in Europe. He wrote to George Heard Hamilton, an old friend and curator of the Yale collection of the Société Anonyme, to express his frustrations:

> It is the American Museums who are responding as if they would not like my work to be seen not only in this country but anywhere outside the country. . . . But you all have one and the same answer, my works are "too fragile." This is one of the slurs which are easily thrown by the critics and also by the Museum Curators; a stigma on my work for which they have no foundation whatsoever except their own fear of the work with which they have no spiritual contact, which they don't understand and which they regard only as another kind of merchandise which they are afraid to handle because they don't know what it is. What proof have you to persist in that common calumny of my work in calling it "fragile"? What is "fragile"? Andrew [Ritchie], in one of his books, giving an account of my work was criticizing them for having no "permenancy" [*sic*]. I never wanted to answer this thoughtless remark of his but I must say now that is the most inexcusable expression ever used by any historian of art about works of art.[12]

Hamilton responded to clarify that the decision not to lend was not his, but he was also supportive of any decisions made that would reduce the threat of harm to Gabo's work. He cited the recent damage to Pevsner's *Portrait of Marcel Duchamp* in the Yale University Art Gallery's collection (discussed in the introduction of this volume), a 1926 sculpture made from cellulose nitrate and metal that was severely broken while on loan in 1956, and needed to be completely remade by the artist. Unfortunately, the new version incorporated different metals than the first, thereby resulting in the rapid deterioration of the plastic and rendering the work unexhibitable. Hamilton defended any decision that would minimize the risk of damage and stressed the importance of protecting the condition of significant collection works so they could be available to students into the future.[13]

Gabo's concerns with his works being labeled fragile or impermanent may have influenced his decision in the 1950s to begin fabrication of newer versions of his earlier plastic works in metal, including *Constructed Head No. 2*. Had he lived long enough, however, to witness the extensive measures taken to understand and preserve his early plastic works, especially in the last few decades, he would have seen the importance accorded to his sculptures within the canon of twentieth-century art. The collective work of scientists and conservators to preserve Gabo's early sculptures has substantially added to the field's ability to care for not only Gabo's artworks but cellulose esters across museum collections too. This body of research, combined with new work on the history and fabrication of the DMA's Ivory Rhodoid *Constructed Head No. 2*, influenced its conservation treatment along with the development of long-term display and storage measures. Although these measures will never stop the slow breakdown of the plastic, they will allow for the continued preservation of this early significant work into the future, for as long as time allows.

Acknowledgments

The author owes deep gratitude to Dr. Steven A. Nash and Charles Wilson for their many contributions, conversations, and memories shared between 2017 and 2023. The author is also most thankful for guidance and support provided by many generous colleagues, most especially Dr. Nicole Myers (chief curatorial and research officer as well as the Barbara Thomas Lemmon Curator of European Art at the DMA), Fran Baas (objects conservator at the DMA), Laura Eva Hartman (paintings conservator at the DMA), Christine Burger (research assistant at the DMA), Hilde Nelson (curatorial assistant at the DMA), and Claire Taggart (conservator at the Nasher Sculpture Center). Without them, the research and conservation treatment work outlined in this chapter would not have been possible.

Hide, Feathers, Beads, and Plastics: The Continuum of Care at the National Museum of the American Indian

Susan Heald and Kelly McHugh

We are conservators based at the National Museum of the American Indian (NMAI). The current holdings have their foundation in the collection of the former Museum of the American Indian, which was assembled by George Gustav Heye (1874–1957) over a fifty-year period. At the time of the transfer of the Museum of the American Indian to the Smithsonian in 1989, there were over eight hundred thousand archaeological and historical collections items spanning the Western Hemisphere. Today, with its mission focused on the continuance of culture, the NMAI is expanding its collection through donations, purchases, and commissions with an emphasis on collecting modern and contemporary art. In working toward the continuance of culture, the NMAI's mission also stresses partnership with Native peoples, thereby necessitating collections development strategies along with programmatic efforts that work in collaboration with community representatives on appropriate standards of care, modes of exhibition and interpretation, and the museum's overall operations. The collection is expanding with new and current works by renowned contemporary artists and ephemera from tribal communities. As such, plastics continually enter the collection, requiring specific care and collaborative stewardship to address the conservation challenges they present.

Materials like hides, ceramics, stones, feathers, porcupine quills, and plant materials are in continuous use by Native and Indigenous people. Used to create everyday functional and ceremonial items as well as contemporary works of art, these materials offer an interrelationship between these elements of the natural world and the people who use them. A profound understanding of resource management connected to cultural protocols applied during collection, processing, and use combine to create the items we see in NMAI's collection. Understanding that the physical item sitting on the museum's shelf is imbued with this knowledge and these practices is an important criterion in its care.

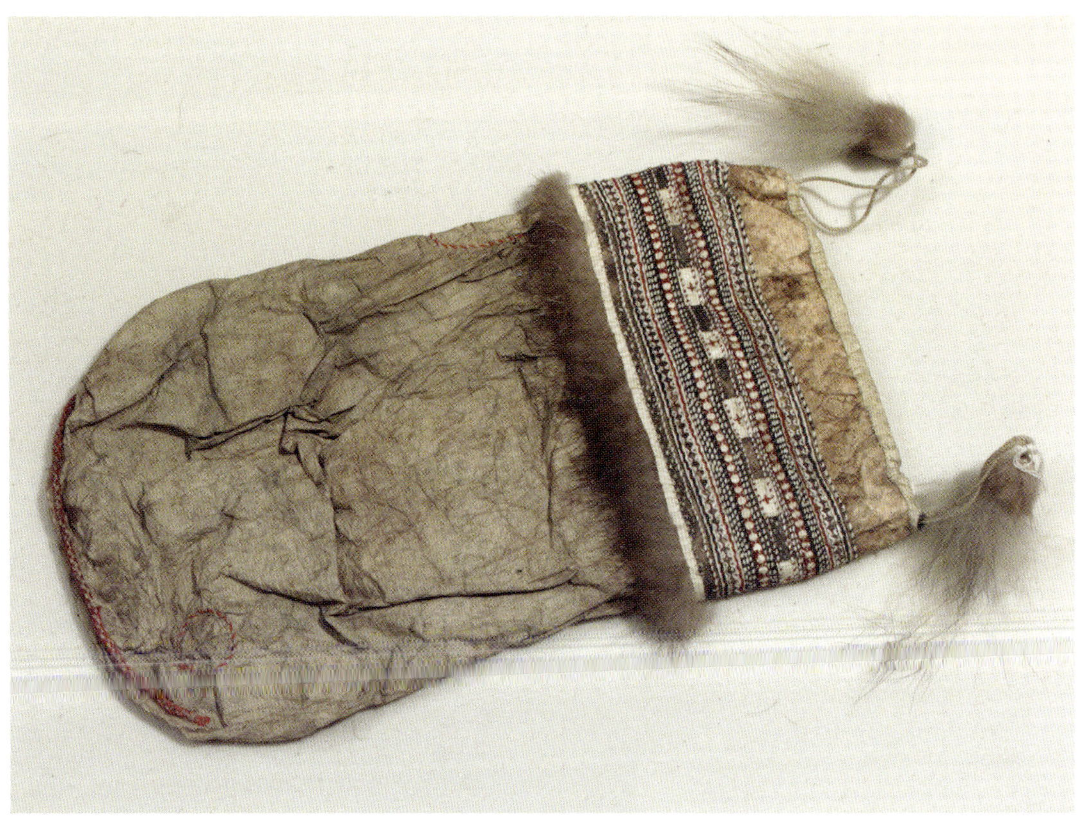

[FIG. 26]
Kuskwogmiut Yup'ik, *Bag*, ca. 1900. Seal gut intestines, fur, and cotton thread, 9.5 × 6 in. (24.5 × 15 cm). National Museum of the American Indian, Smithsonian Institution, Washington, DC, 09/3471.

Functional need has always spurred innovation. An Apache pine pitch basket functioned as a water bottle. Containers created from the intestines of marine mammals by coastal Alaskan people served as the first "plastic bag" [FIG. 26]. It is human nature to adapt and utilize the materials that surround us. Encounters with Europeans introduced a range of materials that were quickly incorporated into daily life. For example, copper metal sheathing from ship hulls used by communities along the Pacific Northwest coast and southeast Alaska became a valued material as well as a symbol of community status and wealth. Glass beads imported to North America from Europe and Asia quickly replaced traditional forms of adornment like dyed porcupine quillwork, which were laborious and time-consuming to craft. A design explosion resulting from the availability of glass seed beads can be seen among tribes like the Apsaalooke and Lakota. As soon as communities had access to European-produced textiles, such as trade wool, velveteen, and silk ribbons, they created and wore beautifully adorned clothing sewn from these fabrics.

The reservation and mission era (1850–87) forcibly removed people from their traditional landscapes and material resources, limiting options for materials used to create items essential for daily and spiritual life. The forced dependency on rationed goods did not extinguish creativity but instead necessitated creative adaptability. Examples can be found in the tinkler cones made from tin cans that created decoration and sound for hide garments, and the flour sacks commonly used as lining fabric, as seen in the eagle feather headdress. In the same way, when plastic was introduced in US society, it was quickly utilized by Indigenous and Native communities.

The first synthetic plastic, Bakelite, was invented in 1907 by Leo Hendrik Baekeland. Its versatility and accessibility proved appealing, especially during the First World War when the scarcity of metal made early plastics a desirable substitute. In 1924, Baekeland was voted "man of the year" by *Time* magazine, which quoted him as saying he was proud to have invented "a non-melting, non-dissolving solid like nothing else found in nature." Tourists to Santa Fe and Albuquerque, New Mexico, were seen wearing Bakelite jewelry in the early 1900s, catching the attention of Pueblo jewelers interested in the new material and expanding their market.

Santo Domingo's Depression Era Jewelry

Santo Domingo Pueblo is situated between Albuquerque and Santa Fe in northern New Mexico, and claims a long tradition of jewelry making. The city is well-known for the intricate use of turquoise, coral, and shell seen in heishi (shell) necklaces and *jaclas* (a pair of beaded earrings that hang on a necklace). Starting in the 1920s, Santo Domingo Pueblo jewelry makers were meshing the nation's oldest Native jewelry-making tradition with cheaper alternative materials, including black plastic salvaged from car batteries and 78 rpm vinyl records as well as colored plastic salvaged from hair combs, kitchenware, and eating utensils. The Albuquerque Woolworth's, opened in 1915, and the Woolworth's on the Santa Fe Plaza, opened in 1935, were common sources of these cheap materials. This access offered jewelry makers a way to dramatically cut costs while retaining the aesthetics of an ancient tradition. Widening the market to railway tourists visiting the region or surrounding Native communities helped bring in desperately needed cash during the Depression. This jewelry was appealing to tourists, and appreciated by fellow Diné (Navajo) and Pueblo residents. Archival photographs show Diné, Cochiti, and San Ildefonso residents wearing these types of necklaces, intrigued by their affordability and vibrant colors. Santo Domingo Pueblo's neighbors embraced the items for personal wear as they were both beautiful and lightweight. This distinctive jewelry was born of necessity and adaptability, utilizing inexpensive, available resources and materials during financially challenging times, thus integrating the past while persevering into the future.

NMAI's focus on the continuance of culture results in the presence of many forms of plastic with a diverse representation of collection items from cellulose acetate hair combs to polycarbonate snow goggles, and includes the San Ildefonso jewelry discussed above along with examples of tribal ephemera [FIG. 27]. More plastic materials are anticipated as the museum's collecting emphasis shifts to contemporary Native art, including contemporary sculpture made from plastics as well as clothing made of natural and synthetic materials, such as modern powwow regalia and couture garments by Indigenous fashion designers. Though NMAI has never undertaken a physical object-by-object survey of plastics in the collection, staff caring for the collection have concerns over the long-term preservation of plastics within the collection. What types of plastics are represented? Do some require cool or cold storage to reduce inherent deterioration? How can we prepare to care for future acquisitions that may include unstable plastics?

In 2020 during the COVID pandemic, museum specialist Makayla Rawlins undertook a virtual plastic survey. Not included in the survey were the archaeological collections, which may contain naturally occurring polymers like rubber and resin, and the archival collections, which include photo and film-based materials such as cellulose acetate. Rawlins used an interactive website, the Plastic Identification Tool, to catalog and describe plastics in the collection by searching descriptions and images in the museum's collection database. The website, developed by an international group of scientists and collections managers with the

[FIG. 27]
(top left) Santo Domingo Pueblo necklace with red plastic inlay, ca. 1970. National Museum of the American Indian, Smithsonian Institution, Washington, DC, 26/5836. (top right) Casino change cup from the Dancing Eagle Casino, Laguna Pueblo. National Museum of the American Indian, Smithsonian Institution, Washington, DC, 25/9684. (bottom left) Hair combs with imitation turquoise, ca. 1950s. Indian Art and Crafts Board. National Museum of the American Indian, Smithsonian Institution, Washington, DC. (bottom right) Alaskan Inuit snow goggles, ca. 1970s. National Museum of the American Indian, Smithsonian Institution, Washington, DC.

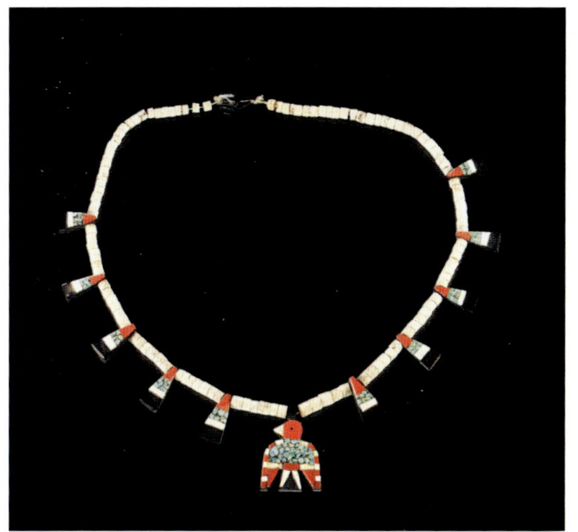

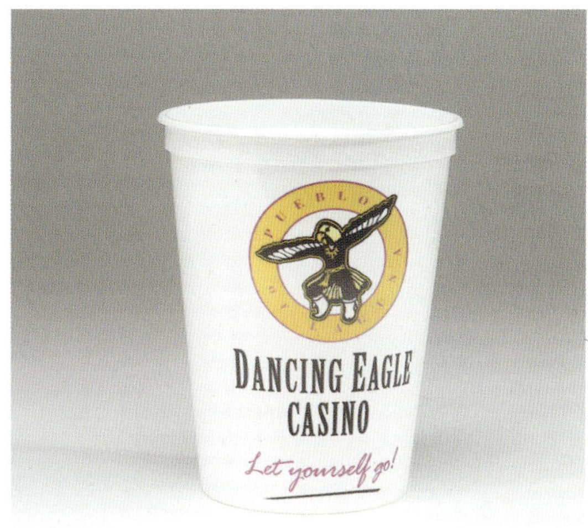
DANCING EAGLE
CASINO
Let yourself go!

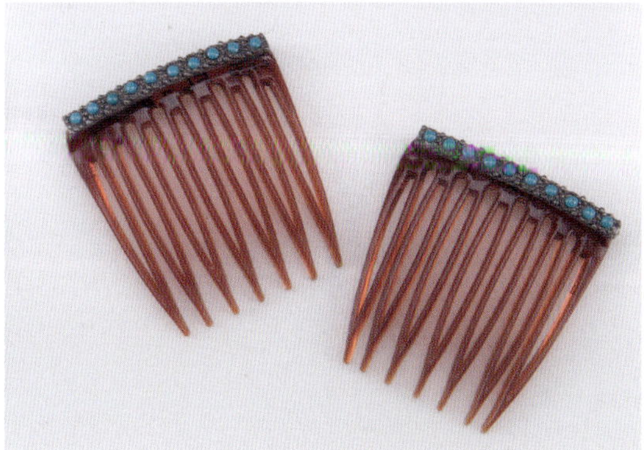

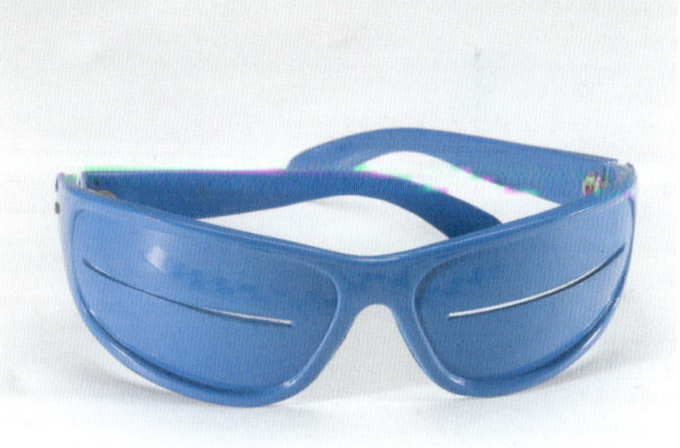

Project Plastics Group is an invaluable resource that uses a creation date and physical characteristics to narrow the possibilities as well as assign a probable polymer identification. While Rawlins's survey method did not catch all plastic items, it provided a baseline for comprehending the scope of plastics in the object-based collection.

This virtual survey identified over five hundred items as "plastic," potentially representing twenty-eight different polymers. The most common polymers are likely polyethylene terephthalate (PET) at 16 percent, acrylonitrile butadiene styrene (ABS) at 14 percent, and polyethylene at 11 percent. In general, problematic plastics like plasticized polyvinyl chloride (PVC) were less common (8 percent), and notoriously unstable plastics such as cellulose acetate and natural rubber appeared uncommon (1 and < 1 percent, respectively). The survey provides a fascinating snapshot of object groupings and some surprising individual items. A group of recent acquisitions includes branded ephemera from tribal casinos and cultural centers, such as water bottles, casino chips, shampoo bottles, and writing pens. Individual items include a package of "Bid Chief" PET-packaged rubber condoms and a beaded polypropylene Pez candy dispenser.

The survey initially missed items not identified as "plastic" in descriptive database fields. One notable example was Brian Jungen's mobile *Crux* made predominantly from plastic luggage, which surprisingly had no plastic-related terms in its description. Other instances include recently collected fashion items where the fiber content and plastic components are not noted. The Plastic Identification Tool includes some polymers used in commercial fiber production such as polyamide nylon and flexible polyurethane, but other commonly used synthetic fibers like rayons and polyesters are not included, so these would not be captured in the survey. Rawlins recommended using standardized terminology for describing plastic items when cataloging future acquisitions so they can be easily identified through a database search.

Powwow Regalia in the Late Twentieth Century

As conservators prepared a twenty-nine-piece men's northern traditional powwow outfit for the 2012 exhibition *Circle of Dance*, the use of an unexpected plastic material was discovered. Artist Robert Tiger Jr. (Hunkpapa Lakota) made the outfit in the 1990s and danced in it many powwows [FIG. 28]. Conservators observed embrittlement along with cracking of the beading support for the vest and choker described as "cotton cloth" in the record. The degradation was not characteristic of the cotton canvas or brain-tanned hide traditionally used as a beading support. Under magnification, it appeared to be a layered composite fabric with an embrittled rubberlike inner layer. After speaking to NMAI

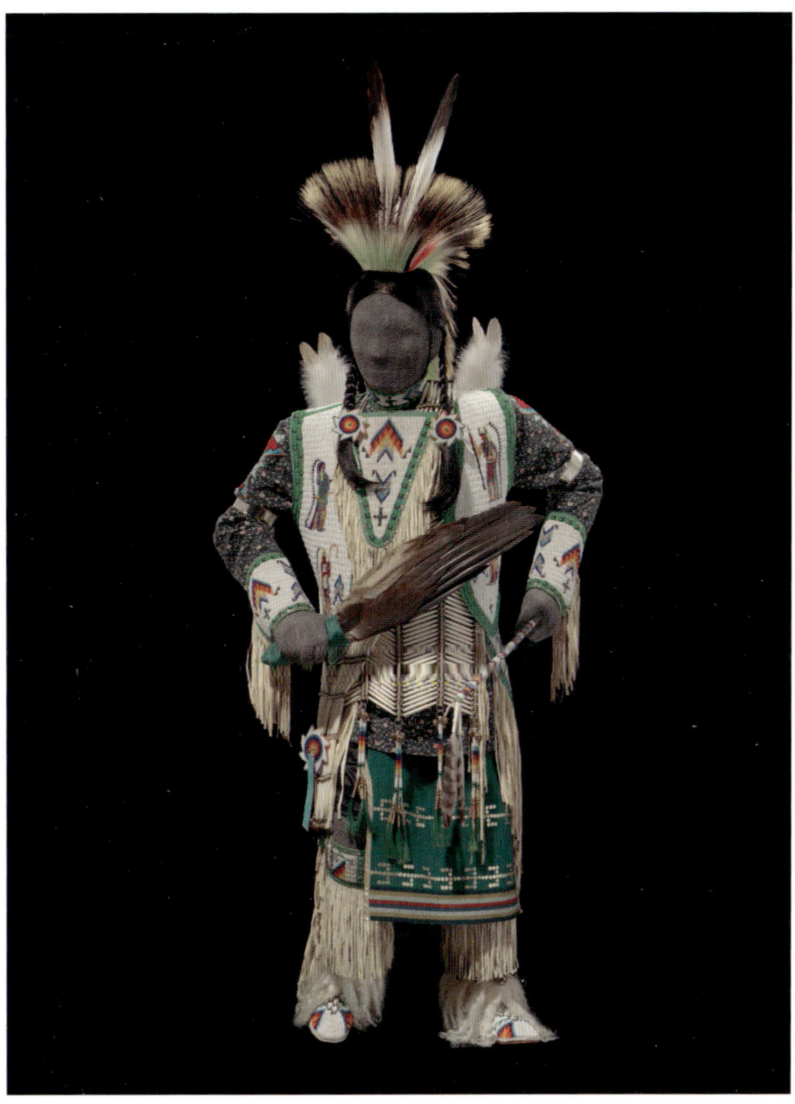

[FIG. 28]
Robert Tiger Jr. (Hunkpapa Lakota), northern-style dance outfit, ca. 2005. National Museum of the American Indian, Smithsonian Institution, Washington, DC.

colleagues and expert beadworkers, curator Emil Her Many Horses (Oglala Lakota) and collections specialist Tom Evans (Skidi Pawnee), as well as artist/regalia maker and former NMAI conservation intern Steve Tamayo (Sicangu Lakota), we learned the material was rubberized flannel—a waterproof fabric comprised of two layers of cotton flannel laminated with an elastomeric "rubberizing" layer sold as baby lap pads and mattress protectors. This realization took conservators down a fascinating historical and analytic research path.

Her Many Horses noted that in the 1960s and 1970s, when the powwow was being revived, some people were working with sinew and semitanned hide, the traditional substrate for heavily beaded pieces, while others were beginning to incorporate new materials. He recalled a piece commissioned by his grandmother that was beaded with nylon monofilament on tanned hide. Alternative materials such as cotton canvas may have been introduced by neighboring communities. Apsaalooke beadworkers had a tradition of using cotton canvas, often recycled canvas tipis, as a beading substrate when the surface was entirely beaded, as Her Many Horses explained in a personal communication in 2013. Both Her Many Horses and Steve Tamayo, also in a personal communication in 2013, said that rubberized flannel rose in popularity in the 1970s and 1980s as an alternative to canvas when the substrate would be entirely covered in beads. Rubberized flannel had several advantageous qualities. Unlike cotton canvas, the inner elastomer layer prevented fraying at the edges. This product could be purchased at fabric stores or retail chains like Walmart and Target. It was economical too; a twin-bed-sized piece of rubberized flannel was one-tenth the cost of a traditional brain-tanned hide, which could be difficult to obtain.

At the time of our research, Tamayo offered some observations as a maker of powwow regalia. Children were being taught to bead using this fabric as a substrate precisely because it was cheap and easily sourced. Tamayo noted that rubberized flannel was widely used for dress tops, aprons, purses, cuffs, side drops, headbands or crowns, braid ties, chokers, bibs, and leggings—all components of a powwow outfit. Steve conveyed that these outfits are made not for longevity but rather the "wow! in powwow." Dancers wear a new outfit every year and sometimes make several outfit changes during one powwow. Tamayo's own daughters produced four to five powwow outfits each year.

We discovered rubberized flannel being used as a substrate for fine art beadwork. NMAI has several pieces by artist Marcus Amerman (Choctaw) that feature beaded portraits on a rubberized fabric substrate, including one of NMAI's founding director, Rick West (Southern Cheyenne), in front of the museum. Amerman confirmed then that he was using this fabric for everything he beaded. He had tried beading on canvas, but preferred the rubberized fabric with the cotton flannel on both sides. In a 2013 conversation in Santa Fe, New Mexico, Amerman observed that his beadwork on rubberized fabric had worn better than his pieces beaded on leather or cotton canvas.

Concern about the composition of the polymers in rubberized flannel and its potential deterioration prompted additional investigation. Because the degraded vest interior could not be sampled for destructive analysis that could confirm the polymer's identity, Rebecca Summerour and colleagues studied five samples of rubberized flannel from their own households as a substitute, representing a fifty-year span of the product. Their analyses revealed different elastomer layers in the various samples: styrene-butadiene rubber, PVC rubber, polyurethane elastomer, and acrylic. Two older fabric samples (from the 1960s and 1990s) contained significant amounts of phthalate plasticizers, including the commonly used and notorious di-ethylhexyl phthalate (DEHP) suspected in a host of health problems such as hormone regulation and cancer. After the risk was acknowledged in the 1980s, DEHP was phased out of products for infants and children—a shift that was observed in the analyzed proxy samples with a later production date. Based on the date of regalia construction, Tiger's outfit is likely constructed of a similar DEHP plasticized PVC. As Thea van Oosten points out in *Properties of Plastics*, factors contributing to the vest's deterioration include plasticizer migration out of the PVC as well as exposure to moisture, heat, and mechanical stress, which could be expected from years of powwow dancing. The layered structure of the flannel over the elastomer, dense beadwork, and fabric linings makes it difficult to assess the plastic's condition on many pieces of powwow regalia. Yet this complex layered construction likely protects the plasticized PVC from deterioration within the museum's controlled environment.

As we prepared for this publication, Tamayo provided updated observations on powwow regalia trends. He noted that rubberized flannel is no longer used because it did not hold up well mechanically with strenuous dancing and was recognized as unstable. Rubberized flannel has also become harder to source as more child- and ecofriendly alternatives come on the market. A stiff synthetic felt is now a preferred beading substrate, and Tamayo says that complex fabric appliqués, adhered and not stitched, are now trending. These new styles will undoubtedly end up in NMAI's collection as powwow regalia trends evolve, and we will need to assess the risk for these synthetic polymers as well.

Artist Brian Jungen and *Crux*

The museum's largest plastic-containing contemporary work of art is undoubtedly Jungen's mobile *Crux* (*as seen from those who sleep on the surface of the earth under the night sky*). The work consists of five suspended animals (a crocodile, shark, possum, emu, and sea eagle) constructed of plastic luggage and steel rods, and a wooden rowboat [FIG. 29]. Jungen fabricated this work on Cockatoo Island for the 2008 Biennale of Sydney using the rowboat to ferry materials to construct the mobile at the installation site. The animals of the mobile are native to Australia and represent constellations that Indigenous Australians used for navigation. The museum purchased this work in 2009 for inclusion in *Strange Comfort*, an exhibition featuring Jungen's work exhibited at NMAI from October 2009 to July 2010. Jungen (Dunne-Za First Nations/Swiss-Canadian), is best known for creating artwork from everyday commercial items. In the 1990s, he disassembled outrageously popular and expensive Nike Air Jordan sneakers, and refashioned them into masks styled after Indigenous Pacific Northwest coast ones, linking the commodification of Nike shoes and Native art. NMAI curator Paul Chaat-Smith (Comanche) states in the exhibition brochure, "Brian Jungen turns objects inside out. By deconstructing them, he changes not only the things themselves, but the way we think about what they used to be, and what they've become. He begins with objects that are useful, ordinary, and comforting. When he's through they are unique, expensive, and useless." Jungen describes his inspiration and creative practices, reflecting that he "likes the intensity of working down to the wire for exhibitions," in Art21's video *Brian Jungen in "Vancouver"* from 2016. During the preparation for *Strange Comfort*, Jungen held a Smithsonian Artist Research Fellowship and was on-site at the museum. Conservation, exhibitions, and collections staff had the good fortune of working with him as his studio assistants as he augmented and changed several of his pieces for the exhibition. The rowboat that had been on the floor at Sydney was inverted and hung from the ceiling to support the mobile animals; the crocodile was lengthened and internally reinforced, and the emu received a refurbished head and legs.

NMAI routinely works in partnership with Native communities to develop exhibitions and at times collaboratively refurbishes collection items for exhibition, but this was the first time the museum had worked so closely with a living contemporary artist. During interviews with conservation and collections staff, Jungen shared directions and insight for the preservation as well as long-term maintenance of *Crux*. The artist was aware that the synthetic materials he uses can deteriorate and supported measures to mitigate damage. As with the powwow regalia, we were not sure which plastics we had in *Crux* and how susceptible they might be to damage. Sampling was not an issue, however, as there was substantial scrap plastic luggage from both the Sydney installation and in-house refurbishment. We originally assumed the suitcases were made of ABS, but it proved a little more complicated. Analysis (FTIR and

[FIG. 29]
Brian Jungen, *Crux (as seen from those who sleep on the surface of the earth under the night sky)*, 2009. National Museum of the American Indian, Smithsonian Institution, Washington, DC, 26/7253.

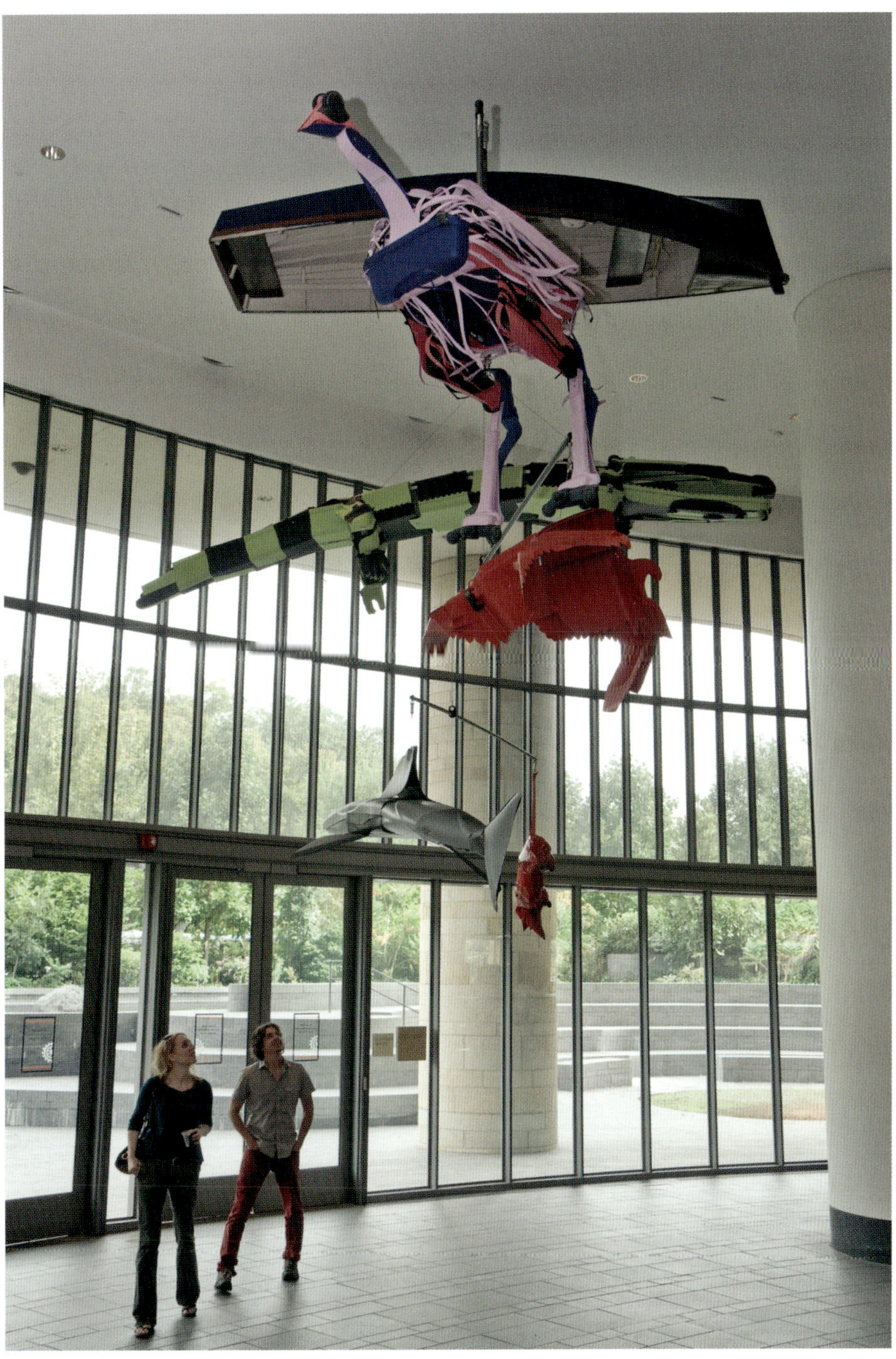

Pyrolysis GC-MS) at the Smithsonian's Museum Conservation Institute revealed the crocodile, emu, and sea eagle luggage were polypropylene or a polypropylene/polyethylene mixture. The possum and shark luggage were composed of ABS and polycarbonate. Though these are considered more stable as plastics go, light damage was a potential threat. All light, particularly UV light, is one of the most damaging factors, introducing polymeric chemical changes along with the fading of dyes and colorants. Recommended light levels for plastics are 50 lux with no more than 75 microwatts per lumen of UV light.

While *Strange Comfort* was to be presented in a black box gallery with controlled lighting, *Crux* was to be suspended, possibly as a permanent fixture, in the museum's rotunda with an oculus overhead and next to a large bank of north-facing windows. *Crux* would welcome visitors and entice them to see the full exhibition upstairs. Light levels from the oculus were in the range of 14,800 lux but swept quickly over the installation site. Light levels from the north windows were twenty-four times higher than the recommended levels sustained over much of the day. While the oculus had UV-blocking film, the north window did not, and UV levels reached 190 microwatts/lumen. These conditions could clearly damage the artwork and shorten *Crux*'s lifespan. To mitigate damage, a protective film was installed on the north windows, blocking 99 percent of the UV light but transmitting 45 percent of visible light. This film made the space safer for *Crux* and maintained the desired aesthetic appearance of the illuminated museum at nighttime—an architectural requirement of its performance.

Preventing the fading of *Crux*'s animals was important to Jungen. To determine how susceptible the plastics might be to color change, samples of the crocodile, emu, and sea eagle were evaluated with microfade testing as well as the on-site light exposure of samples over four to six months with color changes evaluated with a spectrophotometer. The microfade testing suggested all plastic samples performed at the International Organization for Standardization's Blue Wool scale of three or better. The on-site testing, situated in two locations, revealed negligible color change when placed in the north windows and a nonvisible darkening color shift when exposed outside on the director's terrace. The results made us comfortable about exhibiting *Crux* for the run of *Strange Comfort* but not for indefinite display. After the exhibition closed, staff deinstalled and transported *Crux* to reside at the museum's Cultural Resources Center. It is difficult to see the individual animals separated from each other in storage—the magic of the mobile deactivated. In preserving the work, we need to ensure that it is available to the public so as to intrigue and inspire visitors as they enter the museum, and, as the artist intended, provide social commentary.

Care of Plastics in the Collection

Following the virtual survey, Rawlins assigned monitoring levels of one to three based on the type of plastic utilized in the collection items, with one representing stable plastics and three representing those at highest risk of deterioration. Fortunately, most objects can be classified as level one. Approximately 16 percent of the collection might require some monitoring, and approximately 4 percent might need immediate attention. Using the Plastic Identification Tool, we estimate 95 percent of the surveyed items are stable housed under standard museum conditions. Approximately 4 percent of the more vulnerable plastics would benefit from cool storage, while about 1 percent may require cold storage. This was comforting since staff expected a greater percentage of materials needing specialized storage. Plans are underway to conduct a physical survey informed by the virtual one to confirm these findings. This information will be used to design an expanded cool and cold storage area in the NMAI Cultural Resources Center's planned expansion, knowing that plastics will continue to be a fundamental part of NMAI's collection. Additionally, we will implement a practice of accurately identifying the type of plastic in a newly accessioned item at the time of cataloging to aid in developing the most responsible care plan moving forward. The goal is to create a proactive, preventative approach to the care of these collections with accurate identification, appropriate housing, and continued monitoring.

As the use of plastics in society increases and evolves, so does its presence in Native nations. Plastic poker chips, cups, and pens from casinos are now found in NMAI's collection. Author and botanist Robin Wall Kimmerer (Potawatomi) reflects on plastic at the end of the chapter "A Black Ash Basket" in her popular book *Braiding Sweetgrass*. Kimmerer encourages readers to consider the origins of materials in our everyday lives—the oak trees in the floorboards and beach sand in the glass of a jam jar—but for the plastics on her desk, she struggles for connection since plastics are so far removed from the natural world. Though plastics are derived from naturally occurring oil below the earth's surface, the raw material is commercially extracted and chemically polymerized, "creating a vast network of hyper-industrialized goods." The inherent tension of plastic creation and use coupled with the impacts on the earth and people is disproportionally felt in Native America. The continual expansion of fossil fuel infrastructure pollutes the air, water, and soil. The proposed pipeline through the Standing Rock reservation in 2016 was the most visible example of protest. Pipelines, however, are continually being planned through underrepresented communities with little awareness.

The process of petroleum extraction creates environmental havoc, polymer synthesis creates toxic by-products, and the products themselves, like the phthalate plasticizers in rubberized flannel, are toxic. We are now surrounded by plastics in every aspect of our lives, as is reflected in the growing presence of plastic in NMAI's collection. As caretakers, we are aware of the tension between caring for a material whose production causes harm to the communities we serve, and the fact that its accessibility and availability serves both a functional and creative need. The care of plastic items creates new conservation and collections care challenges, but as museum professionals, we will work to understand their composition and preservation needs to ensure their presence endures for future generations.

Plastic Disability Things: Jake's Ankle-Foot Orthoses and the Materials of Disability

Bess Williamson

Plastic is so thoroughly present in the modern world that it cannot be ignored as a cultural material. It is perhaps *the* cultural material of the twentieth century, with little sign of diminished significance. Cultural theorist Roland Barthes wrote in *Mythologies* that plastic was "the stuff of alchemy," its "infinite transformation" the basis of a mythological identity. Design historian Jeffrey Meikle echoed Barthes in naming multiplicity as plastic's core characteristic: "plastic . . . serves so many functions, assumes so many guises, satisfies so many desires, and so quickly recedes into relative invisibility as long as it does its job well." Barthes and Meikle both described plastic's essence as its ability to disappear within that multiplicity of applications and characteristics: both jewelry and buckets (Barthes), both industrial wonder and cheap throwaway (Meikle). Even more than either of these writers explored, plastic is invisible in the microplastic detritus of the ocean garbage patch or toxic traces within all human bodies, as discussed by Jennifer Gabrys in this volume. But few cultural, historical, or artistic commentaries on plastic have looked at its presence in another area of the every day: in the lives of disabled people, many of whom have an outsized relationship to plastic as a material in devices worn daily, on, in, or close to their bodies. Whether in tubing and sterile packaging within the hospital, or devices of bodily support, cleaning, feeding, and other daily activities, plastic "does its job" in intimate, important, and often vexed ways in disabled life.

In the interest of exploring more deeply how plastic operates in embodied human-object relationships, specifically in the relationship between disabled people and their personal devices, we can look more closely at a particular object: an ankle-foot orthotic (AFO) device that was acquired by the Smithsonian's National Museum of American History in 2001 [FIG. 30]. This small boot-shaped device is made of a stiff, molded piece of translucent white plastic with mottled foam padding and red Velcro straps decorated with yellow and orange polka dots. It is under five inches long, fabricated for a small child who still wore newborn-sized clothes at the age of three. This child's orthotic device relates to a number of plastic histories of the twentieth century: its replacement of organic materials (such as leather and metal), shaping into flexible and moldable new forms, and styling as an expression of taste or fashion. As an object of childhood and disability, however, it is set apart somewhat from the image of plastic as the material of rampant consumerism and disposal. Instead, it provides exception to plastic's infinite morphology as it relates to a lack of choice, not unlimited variation.

Katherine Ott, the curator who acquired the AFO for the science and medicine collection of the National Museum of American History, has written about the broad category of "disability things": "the artifacts owned and used by people with disabilities and those that are used upon them or that are encountered in life." As a disability thing, the AFO presents us with plastic's potential as not only an expansively adaptable material but also a part of the medical and caregiving activities in which bodies are assessed and handled by others. For disabled people, especially children and those with intellectual disabilities, consent and choice are limited or actively threatened in medical contexts. All of these categories are relevant to the owner of this AFO, Jake Turley, who wore it from around the ages of three to five. Jake was born in 1996 with a rare chromosomal disorder that produced multiple physical and cognitive disabilities.

Jake's AFO tells me as much about what I *don't* know about Jake as what I do. His mother, Penny Richards, provided all the information that I know about it, typing emails to me while he sat nearby, and she read my questions and responded.[1] While she is open and forthcoming, I know that I am not communicating directly with Jake, and never will. Jake does not speak, and has never stood except supported and in therapy, where his AFOs were used to help him bear weight. In other words, the AFOs are part of the ongoing medical visits that a multiply disabled person experiences: the molding, fitting, and refitting of devices meant to move one's body toward specific physical targets, such as walking, standing, or other movements. Penny asserts several times in our email exchanges that Jake has never liked wearing shoes, which were often an annoyance; the family required multiple trips to orthotists several towns away, and that both parents had to hold Jake for uncomfortable adjustments. While Jake was in school, the AFOs worked as a substitute for shoes and to protect his feet in the jostle of school life. I ask Penny what her or Jake's favorite things about these might have been. She writes that for her, they "felt like armor" as she sent him out into the world, but "he always seemed happier without them," and she doubts he misses them.[2] Given the hassle of getting and putting the AFOs on, Penny does not seem to miss them either.

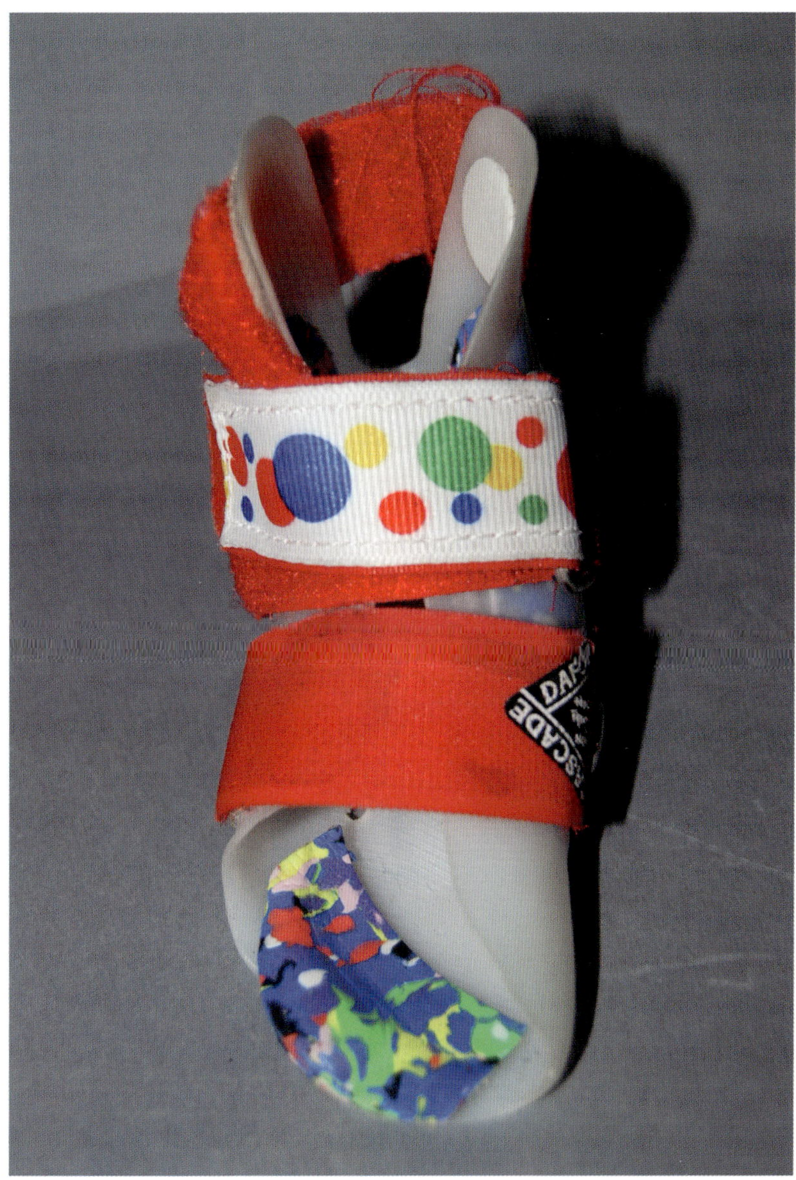

[FIG. 30]
Cascade Prosthetics and Orthotics, Child's orthotic ankle brace, 1999. National Museum of American History, Smithsonian Institution, Washington, DC, 2002.0196.02. Gift of Penny Richards and Peter Turley.

As this chapter will examine, Jake's AFO shows the role of plastic in rehabilitation technology—an industry of prosthetics, orthotics, and assorted other medical as well as consumer products that respond to ideas of disability but not necessarily the ideas of disabled people. Within this slice of material culture, concepts of function and fashion are filtered through the notion of disability as abnormal and in need of correction. Sometimes this correction is physiological, as in reshaping the body to function and look like the body considered nondisabled; sometimes it is social, as in designing the daily tools of disabled life to produce age- or gender-appropriate appearance. Normalizing forces of medical standards and social expectations are present in a long history of disability, including statistical charts and diagnoses identifying abnormality as well as the treatments and technologies prescribed as fixes and cures. For the interests of this chapter, they also have a distinct place in the plastic world. The plastic characteristics of moldability and stylistic variety align with modern discourses around the body and the self that tie together disability, materiality, and technology.

Disability, Materiality, and Metaphor

Jake's AFO offers an object lesson in the medical and social meanings of disability in the late twentieth century (which are for the most part relevant in the present day). Disability studies, an academic field emerging in the last half century informed by disability rights activism, asserts that disability is, like race, gender, and sexual orientation, not an inherent characteristic but rather a social construction with historical and institutional meaning. Disability is a broad term whose meaning is often found in deviation from a norm: medically, in terms of physical or cognitive measures determined as normal, or legally, as an impairment that "limits one or more major life activities," prompting the question of what a major life activity is in a given context. When it comes to the devices associated with disability, they likewise indicate overlapping values of medical intervention and social normality in design for the human body. As artifacts circulating in material culture, devices such as canes, hearing aids, wheelchairs, or AFOs like Jake's encompass historical changes in attitudes about disability and technology. And as personal devices, they also represent the possibilities and limitations of agency for disabled people within medical contexts as well as in society at large.

The AFO, at first look, is clearly a medical object, its white molded plastic and Velcro hook-and-loop straps familiar to many who have worn casts, braces, or splints temporarily or over longer periods. The term *orthosis* derives from the Greek *orthos*, "to correct"; its function is defined by makers as "to stabilize or immobilize, prevent or correct deformity, protect against injury, promote healing, or assist function." In this sense, it reflects what disability studies scholars refer to as the "medical model" of disability: disability as a pathology, usually defined by a diagnosis with assigned treatments and (in some cases) cures. The appearance of the orthosis, however, clearly includes aspects of what disability studies calls "the social model"; first introduced by disabled scholars and activists in the 1960s, this model asserts the meaning of disability beyond the individual diagnosis, given that social and environmental factors contribute to the meaning of one's disabled (or nondisabled) experience too. Jake's AFO is a child's device not only in size but also in style, with its polka dot straps echoing gender-neutral, abstract geometric trends in children's clothing of the 1990s. In addition, social construction can represent the misperceptions of societal norms when applied to products for disabled people. For example, Penny reports to me that later in Jake's life, it was sometimes difficult to find more grown-up-looking fabrics for this growing man whose size still registered as a younger child's.[3]

As numerous disability studies scholars have noted—for instance, Alison Kafer in *Feminist, Queer, Crip*—a binary view of disability models suggests an overly simplified distinction between the embodied "medical" and the seemingly abstract "social." Pain and discomfort are not social constructions, nor are the medical diagnoses of professionals clear-cut or reliably accurate. To separate the medical and social is to overlook the power dynamics of the clinic or the capitalist austerity of modern health care systems, especially in the United States. These models are useful, however, in understanding the inherently multiple functions of these devices, particularly as they traverse the spaces and contexts of disabled people's lives in society. The designs of orthoses and other rehabilitation technologies over centuries have aimed to counterbalance the corrective nature of these devices using materials and styles to reflect typical clothing and expectations of their time. The bright colors and childlike pattern of Jake's AFO might dually suggest making the device friendly and inviting to a child as well as normalizing it within the similar patterns and colors of his and his peers' clothing and toys. In designs for adult devices, aesthetics frequently tend toward modesty or camouflage, further expanding the social role of these devices to diminish the visibility of disability.

Jake's brace is a part of intertwined design and medical histories in which materials and technologies reflect changing demographics and approaches to treatment. The appearance of the brace—quickly identifiable as a child's possession even if we did not know its size—represents nineteenth- and twentieth-century improvements in birth survival rates and pediatric surgery; longer, more active lives for physically disabled children in turn meant a greater need for devices designed for them. These smaller devices joined a historical record of braces, splints, and other bodily supports or prosthetics dating to prehistoric human communities. Medical collections show us precedents in leather, metal, and canvas. Eighteenth-century British poet Lord Byron's leg braces resembled shoes in a dark red leather, with laces up to the ankle and an attachment to align the lower leg [FIG. 31]. Later braces, well through the twentieth century, also used standard boots and shoes, attached to metal braces with leather or fabric straps around the leg. These braces suggest two normative goals—both to straighten limbs and appear as conventional footwear.

The flexibility of plastic, its moldability to the body, exists in tension with the core principle of the orthosis: *to correct*. Orthotics, prosthetics, and the host of other medical devices disabled people adopt under medical care are part of the regime of rehabilitation. A medical term increasingly in use in the early twentieth century, *rehabilitation* implies a return to the "normal" (even if the body in question has never been in that state). Its practices are carried out in medical subfields that were professionalized in the early twentieth century, such as physical therapy, occupational therapy, physiatry, orthopedics, and other offshoots that addressed the ongoing lives of people after acute disease or injury. Disability historian Henri-Jacques Stiker situates rehabilitation as a core principle of the Progressive Era, tied in with other social efforts to reform and make productive those viewed as dependents, including women, immigrants, and the poor. As a turn in the treatment of people with disabilities, rehabilitation, Stiker writes, emerged to reform disabled bodies and minds, thereby making them as close as possible to "ordinary" in terms of appearance or activities such as work. The orthosis does the work of reforming the body by supporting or guiding it into positions considered typical—a normative body shape or "ordinary" looks such as being able to be worn with standard shoes.

In plastic's association with adaptability and change, it shares ground with the technological metaphor of the prosthetic. The prosthesis, a device that replaces a part of the body (as in prosthetic limbs or artificial hearts, eyes, or breasts), has been a dominant object of cultural imagination around technology and modernity since at least the early twentieth century. Modernist architect Le Corbusier imagined modern furniture as "orthopaedic," extending the functions of the human body. Sigmund Freud posited a human being as a "prosthetic god" with modern technologies as "auxiliary organs." The artificiality of mechanical technologies seemed to redefine the body for these modern men, with implications that were in some cases utopian, but other times more grounded. Academic Sarah Jain notes that Freud's own experience with a painful artificial palate that he wore after the surgical removal of cancerous tissue in his throat led to his "simultaneous embodiment of faulty technology and extreme optimism

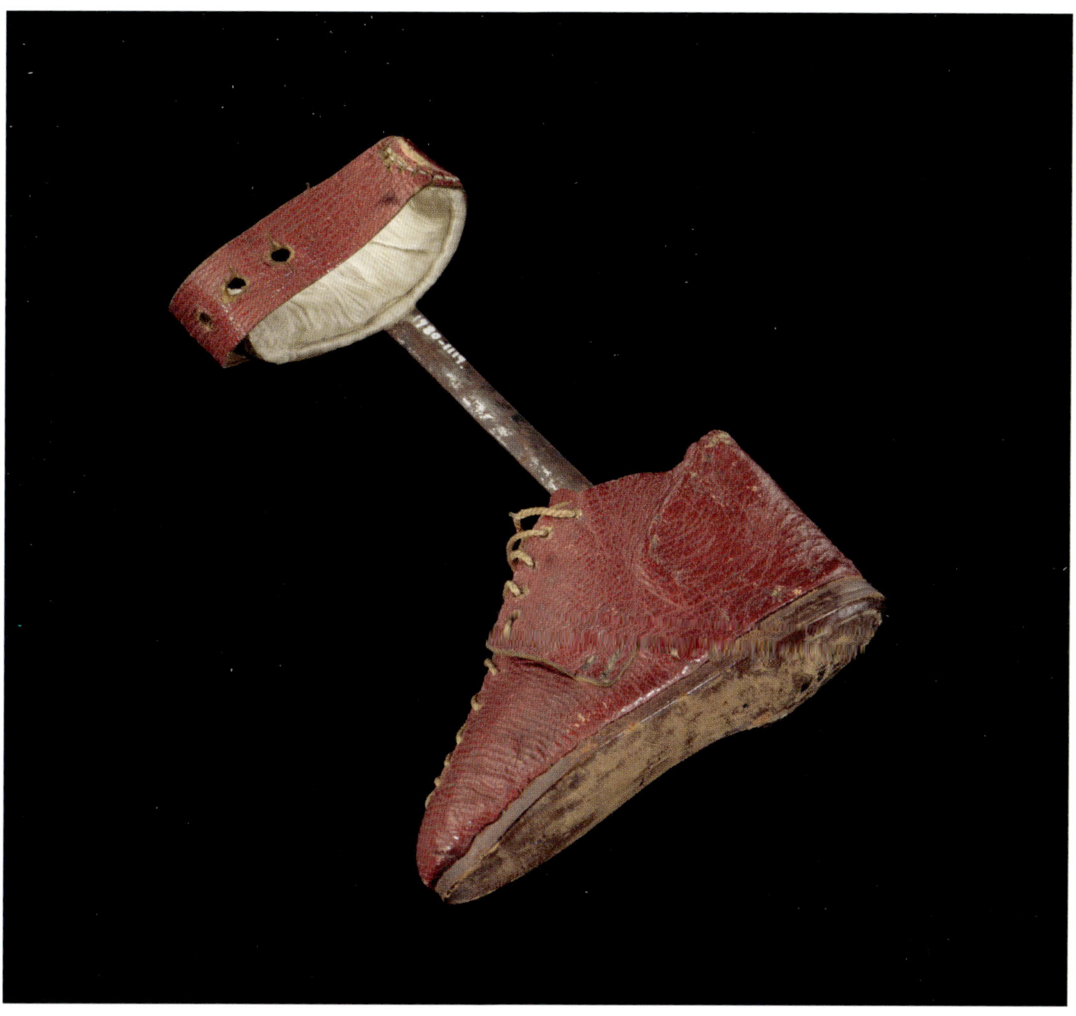

[FIG. 31]
Unknown, Lord Byron's orthopedic boot, England, 1781–1810. Science Museum, London.

about technology's promise." The prosthesis has been taken to represent the technological possibility of extending human function: the camera as a prosthetic eye, and the computer as prosthetic memory or speech. As a number of disabled scholars, especially prosthesis wearers themselves, have pointed out, these accounts tend to oversimplify what a prosthesis is, generalizing the wearer and the device at the same time. Film historian Vivian Sobchack makes the distinction between "*the* prosthetic" as a metaphor and "my prosthetic": "a material but also phenomenologically lived artifact." While the prosthesis has distinct technological agency and performative qualities, Sobchack notes, it "can barely stand on its own and certainly will never go out dancing without me."

Compared to the much-discussed prosthesis, the orthosis takes up little metaphoric space, despite taking a similar role in close proximity to the disabled body. Writing in the 1960s, orthopedist Sidney Licht posited in *Orthotics Etcetera* that the "prosthesis meant both replacement for a part and help for a part," while "orthosis means straightening"—and more expansively, devices "to make life more comfortable, safer and useful." In this broad definition, orthoses are ripe for comparison to modern consumer technologies, particularly those that seek to influence our behavior, like apps that automatically update or even predict what would improve our lives. But the visible prosthetic limb and the possibility of replacement that it seems to represent have kept it clearly in the spotlight of technological metaphors as well as in popular film, fashion and art photography, and literary fiction. By contrast, the category of "orthotics" is broader and more amorphous. The title of Licht's field-establishing tome *Orthotics Etcetera* itself suggests that the category was expansive and hard to limit to a specific set of objects. What is an "orthotics" and what is "etcetera" was not clearly defined, but the book included seventeen chapters on braces and their application, with two chapters on shoes, and several additional sections including wheelchairs, housing, clothing, beds, and cars.

In his inability to clearly define what was orthotics and what was etcetera, Licht inadvertently acknowledged the challenges of categorizing technologies associated with disability. Orthotic devices are just a tip of the iceberg of material things that do not, as Sobchack observed, go out dancing on their own but instead form a part of a varied collection of things useful and fanciful that make up the ingenious art of being disabled.[4] "The line between assistive and prosthetic technology is more like a hyphen," Ott writes in *Artificial Parts, Practical Lives: Modern Histories of Prosthetics*, referring to the many functions of assistance for, or adaptation, replacement, or support of, the body. What these functions mean depends on the disabled person wearing them. This brings us back to Jake, a young man whose communications with the world are often mediated, but who has made it clear to his family that he does not want to wear AFOs (or any grown-up version of them). His orthotics etcetera include socks, a wheelchair, and the caregivers who followed his preference to discard the devices that did not feel comfortable or necessary.

In a recent cowritten piece, "Transmobility: Possibilities in Cyborg (Cripborg) Bodies," three disabled women, Mallory Kay Nelson, Ashley Shew, and Bethany Stevens, reflect on their respective personal equipment and what they call "transmobility," or the "array of options for mobility and movement" that provide an "impetus for creativity and imagination" in their lives. Stevens, for example, writes of feeling a "terrified mess" at the time she adopted a large, powerful electric wheelchair instead of a manual one, but coming to embrace it by naming it Svetlana and "driv[ing] her like my body was always made for her." Disability things come in and out of the lives of these women, self-described "cripborgs" (a portmanteau of the reclaimed "crip" and "cyborg"). Shew, an amputee after osteosarcoma, describes discarding her prosthetic leg at home and gliding across the floor on a wheeled rollator walker, supported by the backward-facing foot that is the distinctive outcome of the type of amputation she underwent. Another preferred stance of mobility is crawling on the floor without any devices at all; this activity is reserved for private home life given that it is the most challenging to typical social norms.

In their multiplicity of material choices, these writers define the varied categories of assistive, prosthetic, and daily technology—or no technology—as changeable. In this sense, perhaps they reveal another aspect of plasticity in their devices, whether they are actually made of plastic or not. Plastic, after all, is a term for a quality of the body itself. Plastic surgery, the practice of shaping the skin, shares its reference to the Greek work *plastikos*, "to mold or form," with the "plastic arts" of sculpture as well as the material to which this volume is dedicated. Sara Hendren in *What Can a Body Do?: How We Meet the Built World*, writes about prosthetic limbs, describes them as highlighting the inherent changeability of the body, the "endlessly plastic and adaptive human body" that wears them. Plastic presents possibilities of change given its capacity to be reshaped, remolded, or simply discarded in favor of a new device, or no device at all.

Ocean Plastic and Orthotics Etcetera

The history of disability things is, in part, a history of the multiplicity of plastics: plastic as moldable against limbs, plastic as clean and sterile in surgery, and plastic as embedded in the body in breast implants or artificial eyes. Plastic disability things include durable, molded orthotic devices and lightweight hearing aids, but they also include single-use plastics. Everyday consumer items such as disposable containers for food and medicine, plastic straws and cups, or the mounds of packaging associated with medical and home goods could all be considered within the realm of orthotics etcetera—the products that bolster the independent movements and activities of disabled people.

Disabled people's use of plastic does not often appear in discussions of the Anthropocene and ways to reduce trash outputs. "Human health and environmental health are dealt with separately," writes Jody A. Roberts in his essay "Reflections of an Unrepentant Plastiphobe," pointing to the separation in government and academic approaches to medical and environmental research. Roberts found himself trying to make sense of his own position as a plastic-conscious consumer once his daughter Helena was born with disabilities such that she required a feeding tube along with a seeming mountain of medical supplies made of plastic and packaged in more plastic. For an "unrepentant plastiphobe," Roberts identified the jarring difference between the abstract hope of mitigating the toxic effects of plastic in daily life and very practical reality of the plastic Helena needed to live.

Roberts's reflection, written in 2010, foreshadowed a coming conflict between environmentalist plastiphobes and disability justice advocates. In the later 2010s, environmentally minded educators and policymakers increasingly advocated for policies to reduce consumer plastic use, whether in plastic bag fees or bans on single-use, disposable plastics—specifically plastic straws. In the late 2010s, plastic straws came into the spotlight as a newly specific target of antiplastic sentiment. Social media videos of sea turtles with straws painfully lodged in their noses provided an emotionally wrenching argument against the widespread use of disposable straws, while news outlets frequently quoted mind-blowing statistics of three to five hundred million straws being used each day. Straws are not easily recyclable, and are hard to segregate from other trash given how light and small they are. For many, this was a design problem that could be solved with a design solution; a host of new products arrived on shelves, from strawless "adult sippy cup" lids to a plethora of aluminum and silicone reusable straws sold at ecoconscious retailers.

As straw bans and café signs exhorting customers to "please don't suck" proliferated, disabled advocates objected that straws were not a superfluous convenience for all consumers. Alice Wong, a leading disabled writer who has a neuromuscular disability that limits her hand and arm movement as well as strength, wrote on the food website Eater that plastic straws and lids allow her to eat independently at her favorite local taqueria or coffee shop. Other people with disabilities echoed Wong's

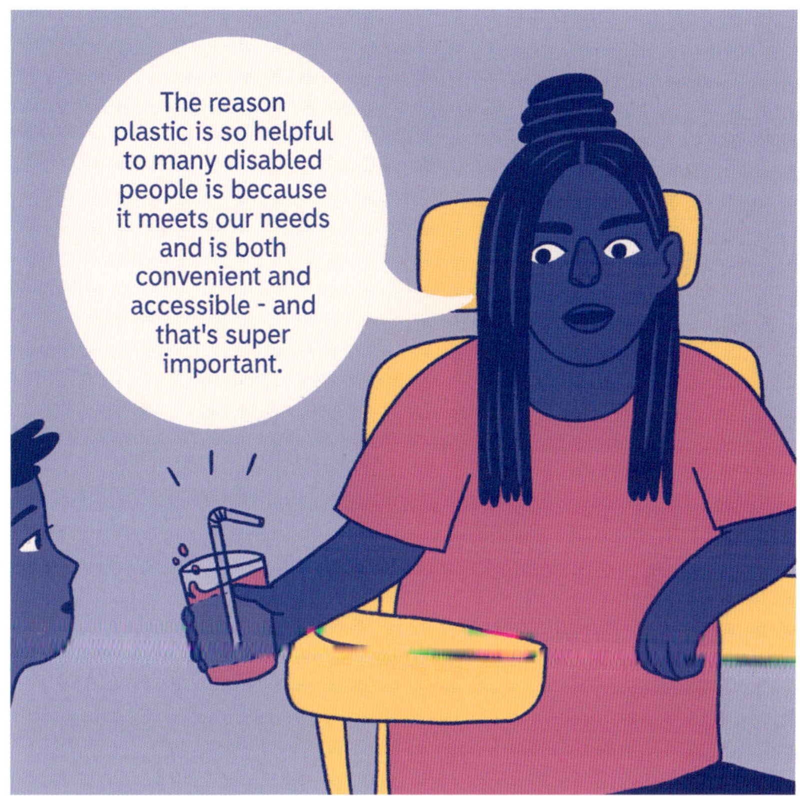

[FIG. 32]
Ananya Rao-Middleton, panel from "What I Wish Non-Disabled People Understood about Disability and Plastic," Greenpeace, March 29, 2022.

complaint, joining her Twitter hashtag #SuckItAbleism as a retort to the virtuous claims of no-straw coffee shops. This was a conversation that specifically called on the material qualities of plastic. As Wong and others pointed out, alternatives like compostable paper straws break down quickly, especially in hot liquids; reusable glass or metal straws can cause injury for people with less motor control in their mouths, and are laborious to store and wash. In a graphic essay for Greenpeace, Ananya Rao-Middleton illustrated this argument, showing a disabled young woman drinking from a plastic cup with a brightly colored straw [FIG. 32]. Rao-Middleton also pointed out that all of this attention to straws overlooked the bigger picture. In reality, disabled people are more likely to bear the brunt of climate change and ecological disaster—a reality that was occurring at the very same time when forest fires displaced and threatened the health of many disabled people in the same West Coast cities that were banning straws.

Plastic straws are a highly visible example of single-use plastic that environmentalists depict as superfluous, but their significance blurs when examined within larger systems of pollution and waste. Roberts's essay seeking a way of reconciling his own preferences with the needs of his daughter reveals the trap of individual consumer choice as an environmental solution. While one's individual choices may mitigate exposure in some small way—even Roberts admits, these are acts against plastic as a "placeholder for the questions and uncertainties about the chemicals in and of our modern world"—the largest polluters do so on scales far beyond those of any individual or household. In their book *Pollution Is Colonialism*, Max Liboiron describes the problematic "we" in many environmental campaigns, such as the assumption that "we are trashing the planet" in equal ways across populations. This kind of universalism flattens differences among communities, such as colonizer and colonized, who have different relationships to land as well as vulnerable populations such as disabled people and corporations that are the largest users of plastic packaging. Liboiron includes Wong and #SuckItAbleism in their analysis: "DuPont has different obligations to plastic pollution than someone with a disability who uses a straw to drink." And in keeping with both Indigenous teachings on a shared cosmology and disability justice principles of attention to the most marginalized, Liboiron asserts that "there can be solidarity without a We." Or as Wong observes, "We should recognize that different needs require different solutions. I'm not a monster for using plastic straws or other plastic items that allow me to live, such as oxygen tubes."

The debates over straws and plastic pollution bring us back to Jake's AFO along with its simultaneous reminders of choice and lack of choice in these disability things. Shape, size, and color in the child's AFO all relate to the possibilities of plastic for this type of device. Still, we know this device to be a rejected one, a part of Jake's rehabilitation treatment that he never enjoyed and ultimately chose to eliminate. For others, an AFO, prosthetic limb, hearing aid, wheelchair, or any range of plastic zip ties or Velcro add-ons to everyday tools may make up an arsenal of choices that a disabled person may use regularly, irregularly, or not at all, perhaps moving between different options in transmobility. Likewise, straws are a part of consumer culture rarely considered in relation to disabled consumers before they were banned; in their restriction, they became newly identified as disability things.

Plastic disability things contain many of the same pleasures, contradictions, and mythologies of any other plastic thing in our modern culture. But in their alchemical qualities of morph and flex, their ability to be molded, colored, or cut and assembled to the particular body and needs, they come to play a key role in disabled people's daily management of their material lives. Whether in the child's AFO, a smoothly gliding rollator, or a plastic straw, plastic does its job, as Meikle suggested above, but it is less likely to disappear, unconsidered. The state of technological things in the lives of disabled people is, in the preferred terms of bioethicist Rosemarie Garland-Thomson, a "misfit": an "incongruent relationship between two things" in which problems arise "not in either of the two things but rather in their juxtaposition, the awkward attempt to fit them together." Plastic played—and plays—a crucial role as it is often either the misfit piece or tool meant to better fit things together. In both roles, it shapes the experience of a disabled person who might slide, plastically, into its hold or resist its presence.

Detecting Particle Ecologies on a Plasticized Planet

Jennifer Gabrys

Numerous plastic particles are present in the air we breathe, food we eat, and water we drink. They are not visible, having broken down into frequently undetectable specks. On average, up to 52,000 such particles circulate through our bodies yearly through our food. This data was published by a group of scientists in a peer-reviewed article, "Human Consumption of Microplastics." In this study, the scientists also note that the number of plastic particles increases from 74,000 to 121,000 particles annually when including the inhalation of plastics. Plastic travels in food and churns through local atmospheres. Particles in tap water and plastic bottles enter bodies, bloodstreams, and organs with water from plastic bottles often containing significantly higher levels of particles. In addition to bodies, plastics travel through oceans and seas, lakes and rivers, soil and beaches, upper atmospheres, and underground aquifers. The planet has plasticized. This chapter, which provides a brief overview of the ubiquity and problem of microplastics, is accompanied by images illustrating microplastics, which indicate the complexity of detecting and sensing plastics. For example, the Alliance for the Chesapeake Bay circulates photographs that make palpable the presence of microplastics in local waters [FIG. 33].

Plastics feature in discussions of the Anthropocene that describe these planetary transformations. Early proposals for this new geological epoch pointed to increasing carbon, accelerating materials production, and rising populations as indicators of how human activity reshapes the planet on a scale equivalent to ice ages or volcanoes. Many critics have rightly taken issue with the conceptualization of the human and agency that the Anthropocene designates. Other versions of -cenes have proliferated to describe the material and ecosocial transformation of the planet, since neither the responsibility nor the burden of this environmental pollution is distributed equally. At the same time, plastics are so pervasive in remaking environments that the term *Plastocene* has been coined to

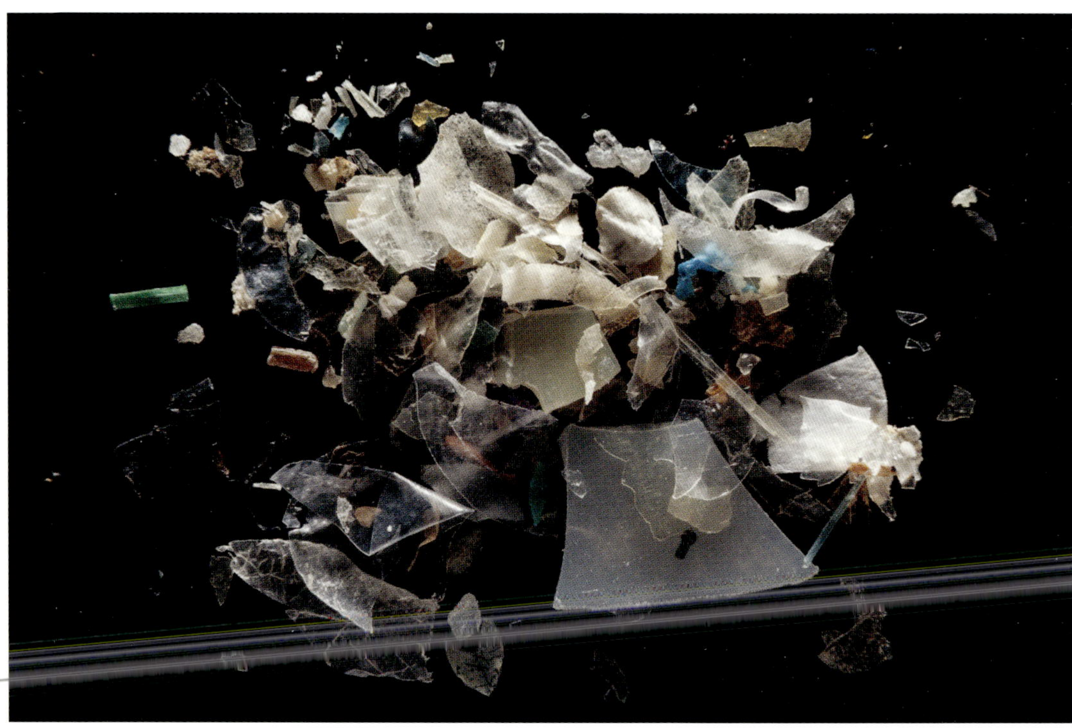

[FIG. 33]
Photo of sample collected by Lance T. Yonkos, Elizabeth A. Friedel, Ana C. Perez-Reyes, Sutapa Ghosal, and Courtney D. Arthur for their study of microplastic presence and abundance in four tidal tributaries to the northern Chesapeake Bay, “Microplastics in Four Estuarine Rivers in the Chesapeake Bay, USA,” *Environmental Science and Technology* 48 (2014): 14195–202.

capture this particular form of earth making. Plastics are yet another indicator of how humans transform the planet through enduring and toxic materials. In this epochal designation, humans are less deliberate makers of objects and more the unwitting recipients of the environmental contamination they have unleashed. As another potential indicator of the Anthropocene, plastics are composed of fossil fuels, as discussed within other chapters in this volume, and rely on typically fossil-fueled energy for their production. The Pleistocene is not an age of deliberate earthworks so much as the clogged and mucky return of the unleashed effects of these plastic material ecologies.

The plasticized planet is composed not of a surface wrapper of cling film or thick mat of identifiable objects; rather, it is a drifting and diffuse debris of plastic particles. The material culture of plastic often unfolds through histories that chart the rise and proliferation of this substance from its initial manufacturing boom in the 1950s. Here, through the shape-shifting object of the particle, I instead consider how future environments are in the making, transforming with and through plastics as they break down, are ingested, seep into waterways, and are hurled through airways to remake planetary conditions. While the image from [illegible] makes evident the ubiquity of microplastics through an aestheticized explosion of their physical presence, scientists also attempt to chart the proliferation of plastics in daily life and how they enter into the environment [FIG. 34]. On this map created by a cartographer with the Marine Litter Vital Graphics project, the designer presents a schematized urban and industrial landscape bordered by waterways, and offers close-up views of the variety of materials—from medical products to agriculture and transportation networks—that generate microplastics. The illustration employs cartographic means to make evident the microscopic world of "particulate emissions," the emission of dust and fiber along with the everyday wear-and-tear of plastic products—a process that makes it so that plastics form the atmosphere that you breathe. The particle, even more than a discrete or recognizable plastic object, is the material that shapes and informs the Anthropocene in the making. In the next section, I consider the particle as an overlooked but crucial object within the material account of plastics. The particle is typically neither stable nor easily detectable. Instead, it is an object that continually transforms and, in the process, alters other entities and environments with which it comes into contact.

Plastics in Process

It can be difficult to conjure an image of plastic particles when media representations of plastic pollution so often focus on identifiable *macroplastics*, which still largely resemble the packaging and consumer

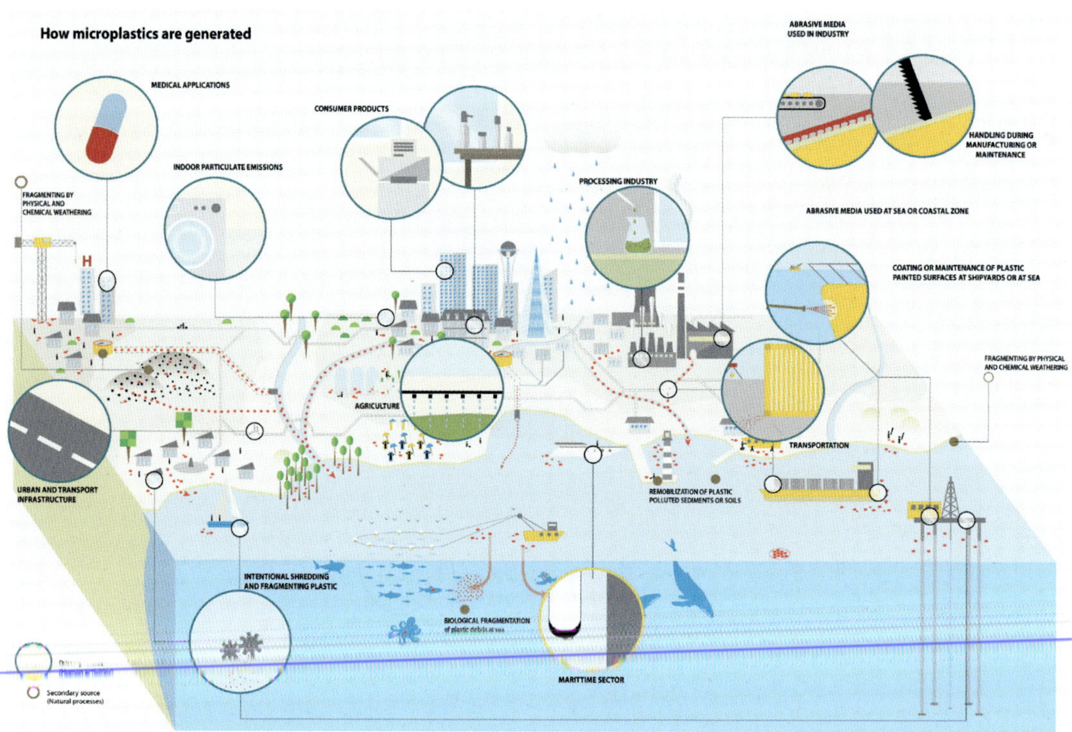

[FIG. 34]
Maphoto/Riccardo Pravettoni, "How Microplastics Are Generated," 2018. Marine Litter Vital Graphics.

goods from which they originate. In plastics research, however, a designation across plastic sizes has been made to attend to plastics as they degrade from larger to smaller and subvisible materials. While definitions differ, macroplastics are objects larger than 5 millimeters, while *microplastics* are objects smaller than 5 millimeters but larger than 0.1 micrometers in size. Still smaller, nanoplastics are less than 0.1 micrometers in size, and like microplastics, are easily ingested or inhaled by humans as well as many other organisms. Some plastic particles start as feedstock for manufacturing processes. Such pellets or "nurdles" are tiny beads that are melted down to form other plastic objects. They often escape, however, from manufacturing sites to enter soil, waterways, and oceans, forming a layer of colored synthetic sand that enters food chains, ingested by marine animals. Microplastics also form through the breakdown of macroplastics, whether through water bottles chipping away in landfills and gutters, or synthetic fabrics releasing fibers through washing and wear. Moreover, once circulating in environments, microplastics can degrade even further while becoming host to microscopic sea life, microbes, and other critters looking for a structure to inhabit. They can also become nanoplastics, even smaller particles that pass more easily into bodies and bloodstreams, taking up residence in organs and remaking biochemical interactions, often to uncertain effects.

Despite the deceptively small sizes of particles, plastics production—and consumption—has grown apace since the 1950s when plastic production was around 2 million tonnes annually. By 2015, it had grown to 380 billion tonnes per year. In their study of all plastics ever produced, Ronald Geyer, Jenna R. Jambeck, and Kara Lavender Law estimated 8.3 billion tonnes of plastics were produced from 1950 to 2017. Of this quantity, 2.5 billion tonnes of plastics were estimated to still be in use, with the remainder discarded. Only 9 percent of discarded plastics have been recycled, while 12 percent have been incinerated. The remaining 79 percent circulate in landfills or environments, with the edges between these sites often being a zone of ongoing transfer since plastics in landfills typically circulate to waterways and seas. While environments and bodies are already awash with accumulating plastics, the condition is only meant to intensify. These same authors estimate that by 2050 the total amount of plastics produced is set to increase to 34 billion tonnes.

Even as plastic production, consumption, and disposal increase, and appeals are made about addressing the problem of a planet "choking" on plastic, many plastics are curiously "missing" from view. This could be because they have drifted to remote environments, especially within oceans, where they are out of sight. But this could also be because they have degraded to smaller microplastics and nanoplastics, and are invisible within the environments and bodies through which they circulate.

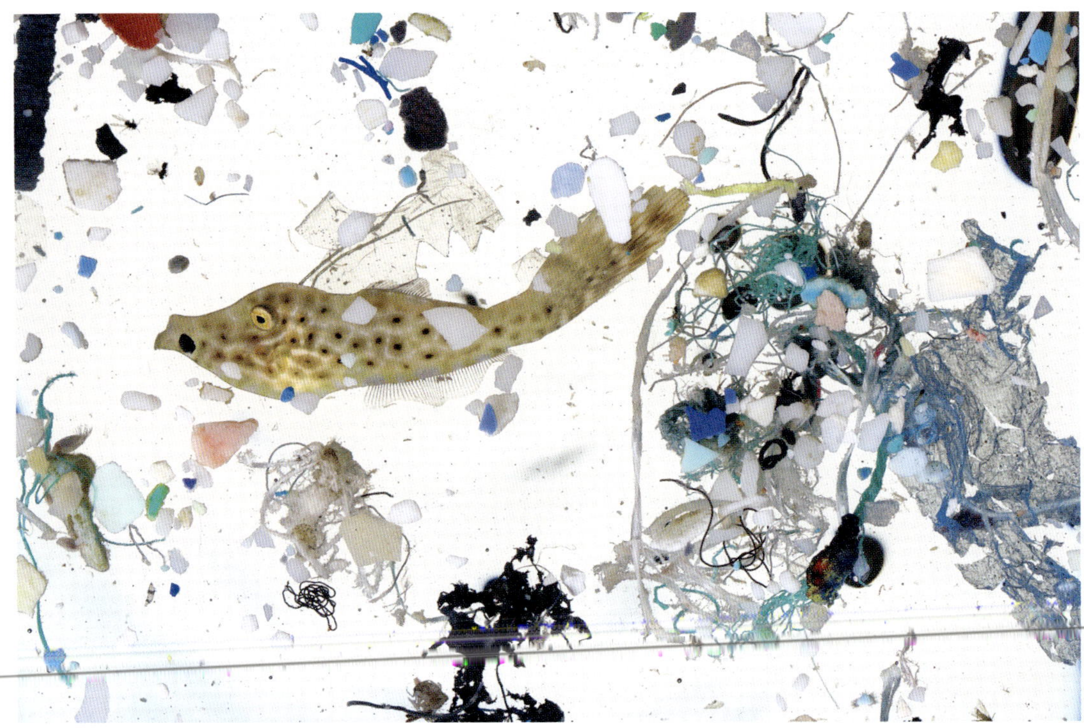

[FIG. 35]
Specimen 5, frame size 5.25 long. Plastic and filefish from plankton sample 799, transect 122, NOAA Cruise MP1812.

Oceanographer and plastics researcher Erik van Sebille refers to this as "dark plastic," similar to dark matter or energy. Such plastics are not easily depicted in images, even those that show plastic breaking down or remaking ecosystems. Such dark plastics are not fully formed consumer objects but instead the missing debris that fragments through plastics in process. Along the way, these elusive particles remake ecologies and particles.

Plastics are a material in process because they are a performative material. Roland Barthes speaks to this condition of plasticity, noting that "more than a substance, plastic is the very idea of its infinite transformation." Yet the plasticity of plastics as matter in motion continues well beyond its useful life in generating uncertain effects in environments and bodies—something detailed at length in the collection of essays titled *Accumulation*. Plastic degrades into heterogeneous particles, fibers, films, and fragments through what medical researcher Elisabeth Gruber and collaborators refer to as "complex pathways involving hydrolysis, mechanical abrasion, thermal degradation, photodegradation, and biodegradation." In these microscopic and more fluid material states, plastic particles pass into organisms and across cell walls, the effects of which are challenging to study and poorly understood. Plastic particles are changeable objects, and therefore continually form new relations with the bodies and environments that they encounter. Plastic particles are airborne, inhaled, digested daily, and intrinsic to the formation of clouds and weather; they travel in bloodstreams, and circulate at the threshold of detection and beyond. Distinct ecologies materialize as particles pass through bodies, environments, and into future planetary effects.

Plastic Bodies

As discussed earlier, it can be unnerving to consider that our lungs are host to the residue of plastic packaging and pipes, tire wear, and plastic bottles. But this is what Lauren Jenner and colleagues found when sampling healthy lung tissue. While searching for particles as small as 0.0003 millimeters through spectroscopy, they found plastic particles lodged in the lungs and potentially remaking airways. Plastic particles and fibers pass not only into our lungs but also into bloodstreams, microbiomes, and many other parts of the body. As the effects of plastics on bodies are poorly understood—leading to questions about whether they exacerbate toxicity, and contribute to inflammation, cancer, or hormone disruption—some researchers have asked how it might be possible to avoid "pathological plasticity." One visual study of microplastics invites a closer microscopic examination of ubiquitous matter that is often taken for granted, such as house dust [FIG. 35]. Researchers produce images such as these to make palpable the contents of what you are inhaling, as plastics enter your bloodstream, heart, lungs, and body. Plastic particles become agents of biological and chemical reengineering as they modify diseases while potentially introducing new ones. Nanoplastics introduce the possibility of remaking organisms and their inhabitations at a cellular level, where plastics can damage neurological systems and lead to reproductive abnormalities. At the same time, novel microbes are emerging that can eat oil at spill sites as well as potentially inhabit and consume plastics in oceans. Plastics are modifying bodies while potentially giving rise to new entities and ecologies altogether.

Plastic Geologies

When plastics are disposed of in landfills, they enter the soil, leaching and migrating from landfill sites to broader landscapes. At the same time, plastics are mixed into the soil as part of agricultural processes and escape from industrial production sites. The Anthropocene, in its more conventional sense, can connote the process of looking for plastics as sedimentary components in the geological record. They are so diffuse and changeable, however, that they do not create a legible material signature, and thus frustrate attempts to identify plastics as evidence of the Anthropocene, as geologist Jan Zalasiewicz and coauthors note in a 2016 piece in *Anthropocene*. We could think of plastics as forming particular fossils and new geologies.

They even form new geologies as they wash up on beaches and create plastiglomerates, a term coined by Patricia Corcoran and her collaborators, as discussed by David Joselit in this volume. These novel geological entities form from degraded plastic, sediment, organic debris, and stone fragments to fuse into a strangely futuristic material [FIG. 36]. And yet these more visible fossilizations of macroplastics are unlikely to be the most enduring record of anthropogenic activity since plastic particles as macroplastics and nanoplastics are far more pervasive as well as transformative material substances. The speculative geologies of the Anthropocene are more thoroughly remade at microscopic scales through the drift and contamination of particles that alter microbial life, shift chemical compositions, and generate a new periodic table of plasticized elements.

Plastic Seas

Oceans and seas are the primary catchments in which plastics circulate. Before they arrive, though, they travel through rivers and other waterways to reach oceans, lakes, and seas. Some plastics are discharged directly from ships through fishing nets, litter, and other debris. But most plastics journey to water from land. Marine biologist Richard Thompson explains that eight million tons of plastics enter the seas each year. As they drift

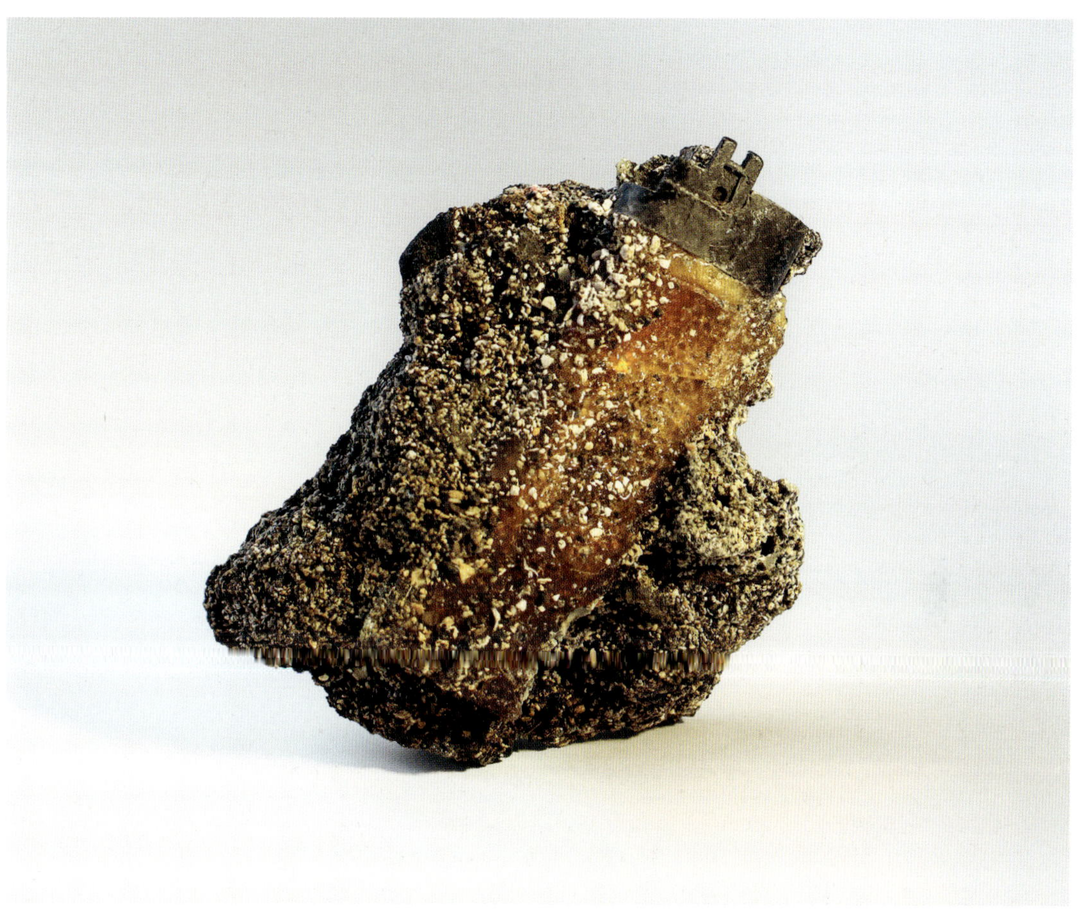

[FIG. 36]
Plastiglomerate sample/ ready-made collected by Patricia Corcoran and Kelly Jazvac, Kamilo Beach, Hawaii, 2012.

and degrade into these planetary sinks, plastics become smaller and suspended in water columns, or descend to deep watery spaces. The Great Pacific Garbage Patch is the iconic figure of ocean plastics, but the image this conjures is often of macroplastics forming a gyre of watery debris. Rather, the patch—and several others like it around the world—is essentially a "soup" of particles drifting within a vast zone of converging oceanic currents, as Charles Moore has noted. These plastic particles do not remain inert or static in these spaces but instead further transform, drift, and sink, becoming vessels, homes, and ingestible substances for many ocean organisms, as all the plastic ever produced likely still exists today in some form. The oceans and seas are a repository of material culture as it transforms into watery particle ecologies.

Plastic Atmospheres

Plastic particles are also adrift in atmospheres. Similar to their circulation on oceanic currents, they oscillate through columns of air and travel varying distances. Plastics drift through skies in London and Shanghai, among many other locations. In cities, atmospheric particulate matter is composed of soot and dust, plastic fibers and debris sloughed off from paint and buildings, and rubber tires and nylon coatings. Plastic particles float to remote areas, such as the French Pyrenees and Arctic landscapes, to become frozen into snowpacks and locked into sea ice. Not only are snowy mountains mixed with plastics, as researcher Melanie Bergmann and authors show, but particles are on the wind and form clouds too. Environmental scientist Mischa Aeschlimann and colleagues found that microplastics and nanoplastics serve the condensation of clouds and ice formation, which could further impact cloud formation. Such effects occur as particles age in environments where the quantity and density of particles lead to a greater or lesser impact on the weather. Local environments with higher levels of microplastics and nanoplastics generate much different weather patterns—from plastic atmospheres to plastic weather—due to the reengineering of the weather that particles can cause.

Plastic Futures

While plastics conjure images of water bottles and appliances, cling film and electronic goods, they are just as present in our bodies and environments as particles. The world of microplastics and nanoplastics remains largely undetectable. Scientists, photographers, and artists use visualization techniques, from maps to microscopic imaging, to make the world of microplastics and nanoplastics evident. Whether drifting into the environment as errant feedstock, or degrading into microplastics or nanoplastics, these particles, fibers, films, and fragments are the residue of our material cultures that remake environments and bodies. The

Anthropocene becomes the Plastocene, less through a singular moment of climatic breakdown or legible anthropogenic event, and more through a slow creeping remaking of the planet at the level of the subvisible. The presentism that plastic disposability enables, as research professor Gay Hawkins notes, unfolds into much longer and menacing timescales. The images accompanying this discussion invite readers to rethink plastics imagery. A shattered water bottle or accumulation of ocean macroplastics is not the only or primary way that plastics circulate. By looking at the world from these more microscopic registers, where images are generated through technologies and devices that are also dependent on plastics, the bodily and ecological milieus that would seem to be free from plastics are already churning with less detectable particles.

At the same time, plastic production continues to rise, exacerbating ongoing climate change by using and burning fossil fuels, which are the material composition of plastics. Shale gas extraction—or fracking—contributes to the rise of fossil fuels for plastics production, which inundates environments with carbon and plastics pollution. Ever new and expanded uses for plastics are invented, from health care to consumer packaging, fueled by the conversion of fossil fuels into plastics. The increase in plastic production and factories is such that a 40 percent rise in plastic production in the next decade alone is projected, leading to what scientist Roland Geyer and authors call the "near-permanent contamination of the natural environment" through pervasive plasticization. Researchers such as Alice Mah have remarked how such developments amplify environmental justice problems. Fossil fuel production and plastics plants continue to be colocated in communities, such as the US Gulf of Mexico and poorer communities in Asia, where the industries' worst environmental impacts occur. For this reason, anthropologist Tridibesh Dey and sociologist Mike Michael engage with the "multiple cares of plastics," where the multiplicity of plastic requires attending to the complex social relations it underwrites.

Perhaps paradoxically, numerous researchers argue that a transition to more sustainable environmental practices is not possible without plastics. From renewable energy technologies to insulation materials, from packaging to prolong food freshness to ensuring transportation efficiency, plastic is positioned as a material central to addressing the climate crisis. Scholars such as Tobias Nielsen scrutinize how plastics have become so pervasive and the mechanisms that allow them to continue to be used so profligately. In the thick of the Plastocene, as it atomizes into an infinite number of particles, we might pause to consider how to remake these material flows. Such a transformation of our material ecologies and cultures could be less dependent on fossil fuels in order to prevent plastics from entering environments, and stop the wayward drift of toxic and uncertain materials. As research scientist Sara Belontz and collaborators maintain, these transformed material environments could also create new ways of working together across disciplines and toward action. Particle ecologies demand that we detect and rework the material relations already remaking us.

Plastic Only: Plastics in the Artist's Studio

Kevin Beasley

Entering into the studio, the natural light from the west-facing windows bleeds into the space, highlighting the shapes and crevices of the warehouse shell, pipes, beams, and fixtures, casting shadows [FIG. 37]. The switches—eight of them—are slick white polycarbonate. The surface is slightly covered in grime from years of fingers, palms, and nails flipping them up and down. The casing is polyvinyl chloride, just as white. Walking into the space, I feel bits and pieces of studio debris, dirt, and dust under my shoes, but it's also slippery. The elasticity of the paint is not enough to provide grip from all the years of sawdust and metal dust. I walk toward the alarm, a plastic rectangular shell too, to punch in the code. The buttons glow luminescent, backlit blue, and depressing them gives both feedback and uncertainty. Their resistance demands a confident input. Sometimes my code is wrong because I didn't push hard enough. The alarm chimes off, and I head to the kitchen, moving around the lumber cart covered with select pine, plywood, fiberboards, Douglas fir two-by-fours, ethafoam sheets, and aluminum flat stock. It is an obstacle course, with sheets and planks obstructing a clear path. It is inventory and comfort. The assurance of material to be cut down, chopped up, and screwed together. Hammered into place to make supports, surfaces, form, and meaning. Jewel cases for objects, thoughts, and ideas.

I enter with a backpack and empty my hands at the kitchen table—a birch plywood top on a simple mild steel square tubular frame [FIG. 38]. With a recently coated polyurethane finish, the table is slick with a raised grain. Textured enough to feel the undulations and its striations, yet glossy enough to clean and maintain. The crumbs and bits from the day before sprinkle across the table, revealing its stale crusty neglect. Sometimes drops of oil, turmeric-tinted sauces, are left to settle. Don't place the bag on the table. It stays on my back while my keys, plastic brass, and steel rattle together on the table. Keys and key fobs are made of all kinds of plastic and rubber. The amount of computer chips within car and gate

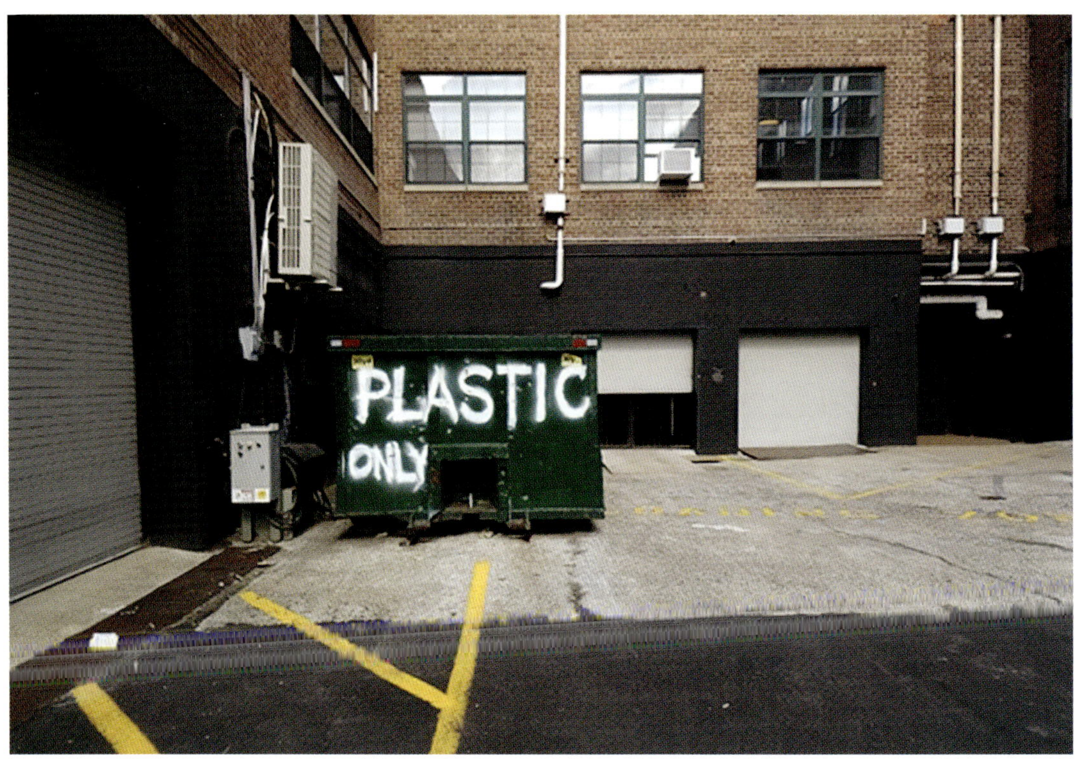

[FIG. 37]
Kevin Beasley,
Untitled, 2023.
Digital image.

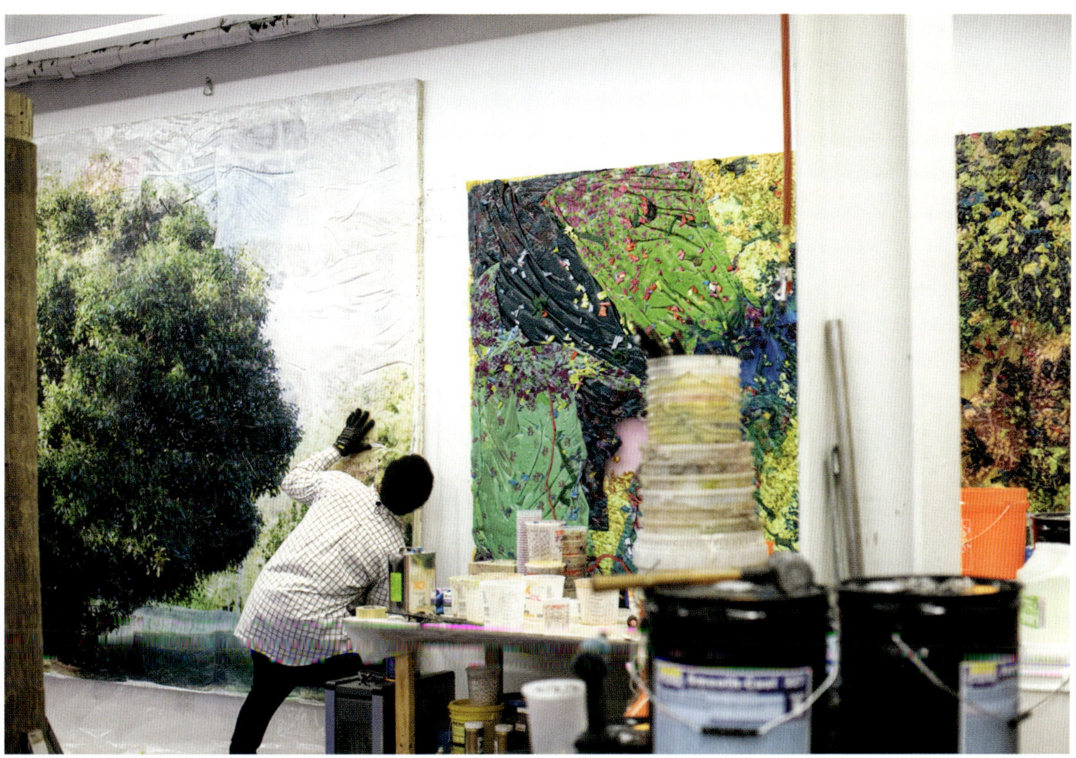

[FIG. 38]
Aundre Larrow, *Kevin Beasley in His Studio*, n.d. Digital image.

keys almost outnumber the house keys. Securing property, owned and leased, borrowed and loaned. The form and shape of keys rest between the fingers, but they really unlock our world that we create and conceal the intimate material of our lives. Photographs, books, clothes, and garments we don't wear anymore; precious jewelry, art, instruments, live animals, and plants too. An architecture that demands the search for more keys, more containers, and more vessels into which we lock ourselves. I carry my keys outside my pocket because there are too many of them. Too much metal and plastic clashing from my belt loop. Pick your pants up.

The stainless steel accents in the kitchen are accompanied by more plastic. The Utilatub slop sink is plastic stained and marred from abrasives. Fast orange in a polyethylene container. Mrs. Meyer's Iowa Pine hand soap, also in plastic. High-density polyethylene (HDPE). Cellophane-wrapped bread and aluminum roasting pans. Plastic utensils from Seamless food delivery, Uber Eats, and Cyclo; Vietnamese cuisine from Long Island City; Tupperware and Rubbermaid. The clear plastic window in the organic almonds paper package, soft and pliable, partially sustainable. A dusty Weber gas grill gets used almost weekly in the summer months, and the plastic knobs are still intact. Its powder-coated steel shape is accented with high-heat plastics. It feels cheap. My shoes squeak as I walk from the kitchen into the main open space.

A glance again over the studio as I head into the thick of it, a dwindled down cotton bale on my left captures the white light, creating a glow amid the stack of containers, trash bins, and miscellaneous furniture. A black injection-molded stackable chair nestles within the pile of containers and the cotton bale. This is where we typically pull cotton from the bales and into clear plastic recycling bags. The process is meditative, but its purpose is to release the fibers of the cotton from the densely packed block that it is ginned into in order to reestablish the bolls—small fluffy balls of fiber and twig shavings. The bales are shipped on wooden pallets because they are five hundred pounds each. They are typically wrapped in a woven polypropylene plastic that is strap-clamped together to help it hold its shape. Polypropylene is typically used for food packaging because it is approved by the Food and Drug Administration. Considered safe, it is also used in many farming packaging containers. The imprint of the Seal of Cotton logo signals its direct link to trade and commerce while also harboring the brand of Cotton Incorporated to signal to the US public its importance to our everyday life.

Mass production in agriculture, like many other industries, relies on heavy petroleum processing in order to produce plastics. The auto industry is another that comes to mind, and the medical industry, in which plastics have increased the sanitary quality of hospitals while becoming an unsustainable haven for single-use plastic waste, as well.

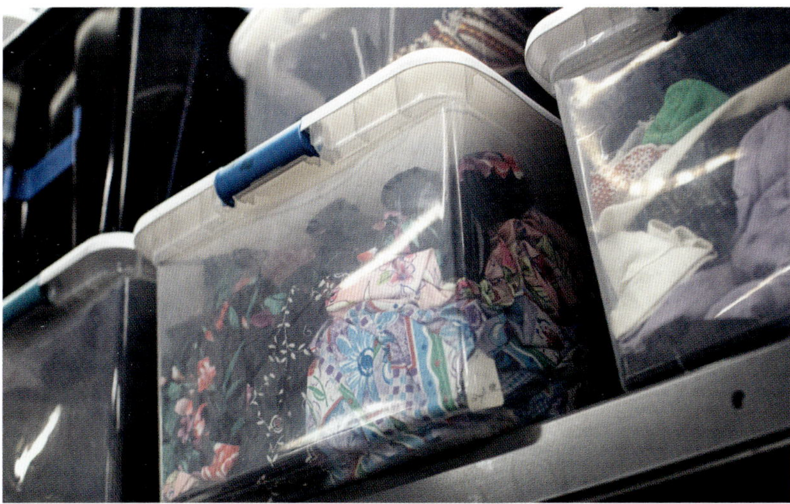

[FIG. 39]
Aundre Larrow, *Untitled*, n.d. Digital image.

Single-use silicone. Single-use nitrile. Single-use latex. Single-use polyester. Single-use polypropylene. Single-use rayon. Single-use white birch tongue depressors. Single-use cotton swabs. Bleached. Single-use isopropyl alcohol whatevers. Stainless steel needles. Niobium tubes, high carbon steel wire; nickel-plated. I don't use needles in my studio; there is no need. But it's interesting to think about the transfer of material, liquids, and goods from one location to another, from one vessel to the next, and into the lifeblood of our aspirations. Especially if one considers the conditions of one location, and how it can dramatically differ from another. The idea of a conduit, which moves materials, thoughts, ideas, and intentions from one place to another. With needles, it can be done surgically. Sometimes these things do not align. They are incongruent. All you can do is embrace the contradiction in order to deliver what is necessary, carefully, to those who seek it. The receiver is different than the deliverer. An empty vessel serves up material in ways that other forms cannot: a bucket, container, pan made of HDPE. The slippery surface allows paint, oils, and creams to slide down easily without much residue. There are buckets everywhere in my studio. I use and reuse them a lot. Caked with pigmented, hardened resin along their edges and corners, they are some of the most important items in my studio.

HDPE is widely used in water bottles, food containers, piping, and buckets. It is one of the most versatile plastics in production. In fact, polyethylene is the most widely used plastic in the world. It occupies every corner of the studio. To the right, blocking a suite of windows, are three large steel shelving units on wheels. Large, clear plastic bins filled with clothing, wires, cables, tools, electronics, and everything in between fill the shelves up to the ceiling, towering above the lights. The sight is overwhelming for those who do not know where these materials come from: unfinished sculptures, half-resined T-shirts, underwear, and tube socks. For me, it is my thought process stacked twelve feet high. So many cotton and wool garments are tucked away, bulging out of the containers from years of accumulation. They are gathered from many loved ones. They are clothes that I would wear in the cold streets of Manhattan or sitting in my car in Detroit waiting for it to warm up, dirtied and bundled away, chapped hands and residual fingerprints. The polyester and cotton blends, the spandex and silk garments, sweaters, scarves, and socks, await to be folded into the mix of a new whatever the fuck. I think about this often. How long a material sits, waiting to be subsumed. Reused. Recycled [FIG. 39].

Latex, an acrylic-based paint, cures to a smooth texture although it degrades over time, like everything. We were told plastic never degrades and therefore is bad for the environment. But I'm also informed that polyurethane plastic dissolves back into a liquid again. Chemistry will still build new polymers that have yet to be conceived. The earth contains all of these possibilities, but only through human experimentation, intellectual thought, and synapses can these chemical compounds be created, pushing the bounds of natural [illegible]—as if screaming in the face of a monstrous black hole as the skin on your face peels away into its void. Fleshy membrane charred by the force of a million suns—hot and black. It fucking hurts sometimes as the nerves electrify, bleed, and fry.

Plastic melts quickly. I remember that for an exhibition, we installed lights in a hanging chandelier-type sculpture, with a speaker array encircling the sculpture's lower half. We needed to fireproof the work. As a test, we burned a sample of the resin-drenched garment, and it went up in flames immediately. Everything eventually goes up in flames without suppression. We never got that back [FIG. 40].

The important thing about touring my studio is that you look. Your body shrinks among the abundance of color, the sharp and soft edges, the masses of grain, liquids, and tumbled garments. The fibrous collections. The hairy surface. Layers of dust accumulation contain fragments of plastic, textiles, and lint. Tools of steel and aluminum rigidly measure the extremities. The elasticity of this space holds cosmic levels of possibilities, and I am only scratching the surface. The prospects are that the eventual outcome is a kind of death that either incinerates our remains or pumps us with formaldehyde; it's just a matter of consciousness. If one claims to be conscientious and observant of what state we are in, fraught with joyous rage, then how can you not resist the periphery view that reveals an event horizon and lean into it like no other? Just make sure to wear a mask so you don't burn from the inside out.

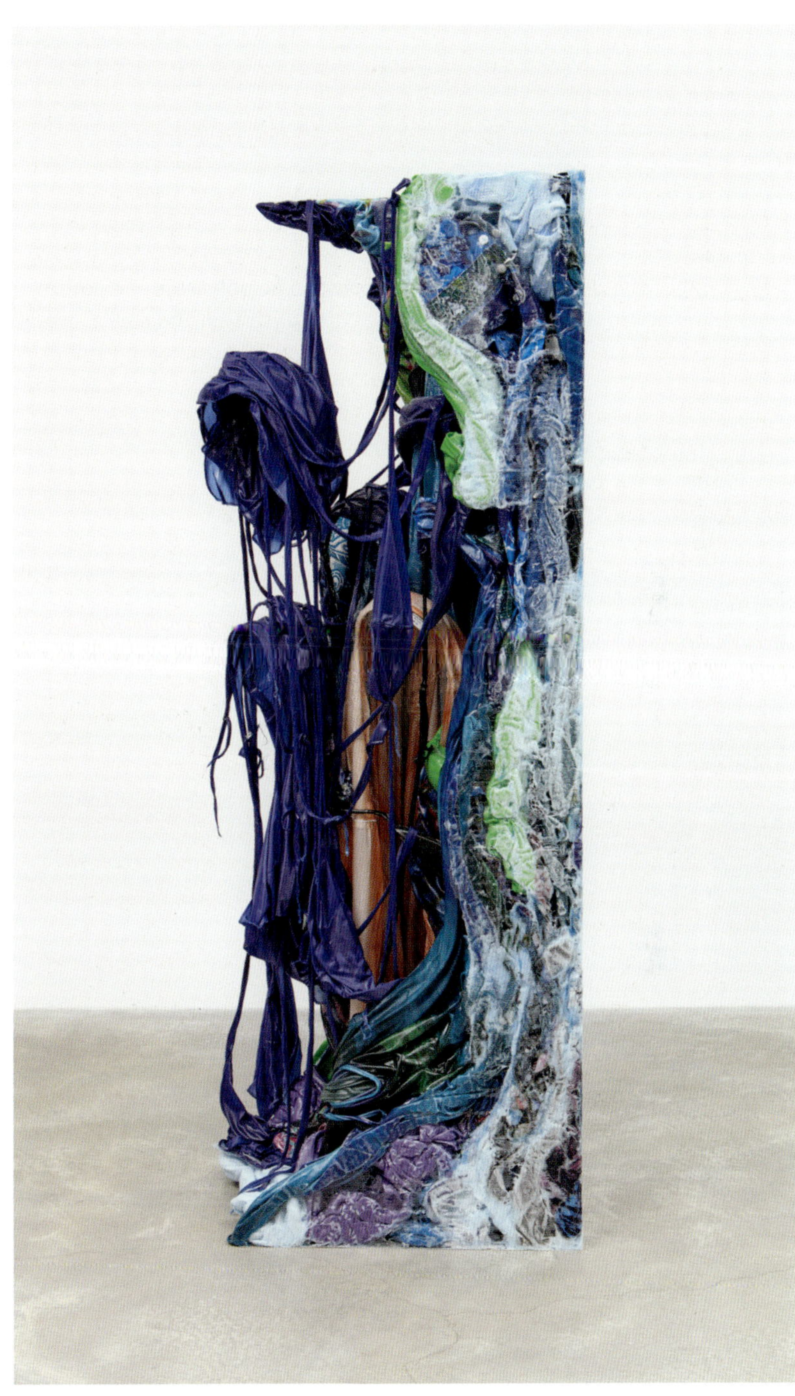

[FIG. 40]
Kevin Beasley, *The Blues (Delta)*, 2018. Polyurethane resin, housedresses, kaftans, t-shirts, du-rags, fishing net, 72 × 20 × 27.5 in. (182.9 × 50.8 × 69.9 cm).

Notes

The Conservation Treatment of a Naum Gabo

1. While ART/WORK volumes in general are aimed at consolidating references under a further reading section, this citational practice is difficult for many conservators who rely on conversations and correspondence. For this reason, Torok asked to include certain citations that were central to her research. Naum Gabo, carbon of letter to James Johnson Sweeney, September 28, 1952, YCAL MSS 541, series 1, box 25, Naum Gabo Papers, Beinecke Rare Book and Manuscript Library, Yale University.
2. Michele Derrick, Dusan Stulik, and Eugena Ordonez, "Deterioration of Cellulose Nitrate Sculptures by Gabo and Pevsner," in *Saving the Twentieth Century: The Conservation of Modern Materials—Proceedings of a Conference Symposium 1991*, ed. D.W. Grattan (Ottawa: Canadian Conservation Institute, 1993), 169–82.
3. Beth Price, Sally Malenka, Ken Sutherland, Andrew Lins, and Janice H. Carlson, "Naum Gabo's Construction in Space: Two Cones: History and Materials," in *Plastics: Looking at the Future and Learning from the Past, Papers from the Conference Held at the Victoria and Albert Museum, London, 23–25 May 2007*, ed. Brenda Keneghan and Louise Egan (London: Archetype Publications Ltd., 2008), 81–88.
4. James Johnson Sweeney, telegram to Naum Gabo, May 3, 1952, YCAL MSS 541, series 1, box 25, Naum Gabo Papers, Beinecke Rare Book and Manuscript Library, Yale University.
5. Naum Gabo, carbon of letter to James Johnson Sweeney, April 5, 1952, YCAL MSS 541, series 1, box 25, Naum Gabo Papers, Beinecke Rare Book and Manuscript Library, Yale University; Virginie Pevsner, letter to Miriam and Naum Gabo, May 8, 1952, TGA 9313/1/1/5/16, Papers of Naum Gabo, Tate Archive, London.
6. Naum Gabo, carbon of letter to James Johnson Sweeney, May 7, 1952, YCAL MSS 541, series 1, box 25, Naum Gabo Papers, Beinecke Book and Manuscript Library, Yale University.
7. Pevsner, letter to Miriam and Naum Gabo.
8. Antoine Pevsner, letter to Naum Gabo, May 18, 1952, TGA 9313/1/1/4/191, Papers of Naum Gabo, Tate Archive, London.
9. Naum Gabo, carbon of letter to Walter Lenox, October 3, 1952, YCAL MSS 541, series 1, box 26, Naum Gabo Papers, Beinecke Rare Book and Manuscript Library, Yale University.
10. Graham Williams and Nina Williams, personal email communication, October 18, 2018; Steven Nash, personal email communication, November 16, 2017.
11. Charles Wilson, personal communication, November 22, 2017.
12. Naum Gabo, carbon of letter to George Heard Hamilton, February 3, 1965, YCAL MSS 541, series 1, box 10, Naum Gabo Papers, Beinecke Rare Book and Manuscript Library, Yale University.
13. George Heard Hamilton, letter to Naum Gabo, February 17, 1965, YCAL MSS 541, series 1, box 10, Naum Gabo Papers, Beinecke Rare Book and Manuscript Library, Yale University.

Plastic Disability Things

1. Penny Richards, "Re: Research Question for You," August 12, 2021; Penny Richards, "Re: Research Question for You," February 7, 2022; Penny Richards, "Re: Research Question for You," April 18, 2022.
2. Richards, "Re: Research Question for You," February 7, 2022.
3. Richards, "Re: Research Question for You," February 7, 2022.
4. I refer here to Neil Marcus's often-quoted statement that "disability is not a brave struggle or courage in the face of adversity. Disability is an art. It's an ingenious way to live." Quoted in Josh A. Halstead, "Disability Theory," in *Extra Bold: A Feminist, Inclusive, Anti-Racist, Nonbinary Field Guide for Graphic Designers*, ed. Ellen Lupton, Farah Kafei, Jennifer Tobias, Josh A. Halstead, Kaleena Sales, Leslie Xia, and Valentina Vergara (New York: Princeton Architectural Press, 2021), 40.

Further Reading

Adams, Rachel, Benjamin Reiss, and David Serlin. *Keywords for Disability Studies*. New York: NYU Press, 2015.

Aeschlimann, Mischa, Guangyu Li, Zamin A. Kanji, and Denise M. Mitrano. "Potential Impacts of Atmospheric Microplastics and Nanoplastics on Cloud Formation Processes." *Nature Geoscience* 15 (2022): 967–75. https://doi.org/10.1038/s41561-022-01051-9.

Balachandran, Sanchita. "Object Lessons: The Politics of Preservation and Museum Building in Western China in the Early Twentieth Century." *International Journal of Cultural Property* 14, no. 1 (2007): 1–32.

Barthes, Roland. "Plastic." In *Mythologies*, translated by Annette Lavers, 97–99. New York: Farrar, Straus and Giroux, 1972. Originally published in 1957.

Belontz, Sara L., Patricia L. Corcoran, Heather Davis, Kathleen A. Hill, Kelly Jazvac, Kirsty Robertson, and Kelly Wood. "Embracing an Interdisciplinary Approach to Plastics Pollution Awareness and Action." *Ambio* 48 (2019): 855–66.

Berger, Michelle E. "A Delicate Balance: Packing, Handling and Installation of Ephemeral Works by Eva Hesse." *Journal of the American Institute for Conservation* 47, no. 1 (2008): 27–40.

Bergmann, Melanie, Sophia Mützel, Sebastian Primpke, Mine B. Tekman, Jürg Trachsel, and Gunnar Gerdts. "White and Wonderful? Microplastics Prevail in Snow from the Alps to the Arctic." *Science Advances* 5, no. 8 (2019): eaax1157.

Boetzkes, Amanda. *Plastic Capitalism: Contemporary Art and the Drive to Waste*. Cambridge, MA: MIT Press, 2019.

Boudia, Soraya, Angela N. H. Creager, Scott Frickel, Emmanuel Henry, Nathalie Jas, Carsten Reinhardt, and Jody A. Roberts. *Residues: Thinking through Chemical Environments*. New Brunswick, NJ: Rutgers University Press, 2021.

Buchloh, Benjamin H. D. "Open Letters, Industrial Poems." *October* 42 (Autumn 1987): 88–89.

Buranyi, Stephen. "The Missing 99%: Why Can't We Find the Vast Majority of Ocean Plastic?" *Guardian*, December 31, 2019. https://www.theguardian.com/us-news/2019/dec/31/ocean-plastic-we-cant-see.

Campt, Tina M. A. *Black Gaze: Artists Changing How We See*. Cambridge, MA: MIT Press, 2021.

Chokshi, Niraj. "How a 9-Year-Old Boy's Statistic Shaped a Debate on Straws." *New York Times*, July 19, 2018, sec. Business. https://www.nytimes.com/2018/07/19/business/plastic-straws-ban-fact-check-nyt.html.

Condie, David N. *An Atlas of Lower Limb Orthotic Practice*. London: Chapman & Hall Medical, 1997.

Corcoran, Patricia L., Charles J. Moore, and Kelly Jazvac. "An Anthropogenic Marker Horizon in the Future Rock Record." *GSA Today* 24, no. 6 (2014): 4–8.

Cox, Kieran D., Garth A. Covernton, Hailey L. Davies, John F. Dower, Francis Juanes, and Sarah E. Dudas. "Human Consumption of Microplastics." *Environmental Science & Technology* 53, no. 12 (2019): 7068–74.

Davis, Heather. *Plastic Matter*. Durham, NC: Duke University Press, 2022.

Derrick, Michele, Dusan Stulik, and Eugena Ordonez. "Deterioration of Cellulose Nitrate Sculptures by Gabo and Pevsner." In *Saving the Twentieth Century: The Conservation of Modern Material—Proceedings of a Conference Symposium 1991*, edited by D.W. Grattan, 169–82. Ottawa: Canadian Conservation Institute, 1993.

Dey, Tridibesh, and Mike Michael. "Caring for the Multiple Cares of Plastics." In *Plastic Legacies:*

Pollution, Persistence, and Politics, edited by Trisia Farrelly, Sy Taffel, and Ian Shaw. Alberta: Athabasca University Press, 2021.

Domingues, Heloisa, Maria Bertol, and Emilie Carreón. "Rubber." In *New World Objects of Knowledge: A Cabinet of Curiosities*, edited by Mark Thurner and Juan Pimentel, 51–56. London: University of London Press, 2021.

Editorial Board. "Notes from the Plasticene Epoch." *New York Times*, June 14, 2014. https://www.nytimes.com/2014/06/15/opinion/sunday/from-ocean-to-beach-tons-of-plastic-pollution.html.

Erickson, Mark St. John. "Williamsburg Exhibit Reviews Folk Jewelry Spawned by the Great Depression." *Daily Press*, July 11, 2015. https://www.dailypress.com/2015/07/11/williamsburg-exhibit-reviews-folk-jewelry-spawned-by-the-great-depression/.

Evans, Mel. *Artwash; Big Oil and the Arts*. London: Pluto Press, 2015.

Finch, Jacqueline. "The Ancient Origins of Prosthetic Medicine." *Lancet* 377, no. 9765 (February 12, 2011): 548–49.

Gabrys, Jennifer. *Digital Rubbish: A Natural History of Electronics*. Ann Arbor: University of Michigan Press, 2011.

———. "Speculating with Organisms in the Plastisphere." In *Pinar Yoldas: An Ecosystem of Excess*, edited by Heike Catherina Mertens, 48–59. Berlin: Ernst Schering Foundation, 2014.

Gabrys, Jennifer, Gay Hawkins, and Mike Michael, eds. *Accumulation: The Material Politics of Plastic*. London: Routledge, 2013.

Geyer, Roland, Jenna R. Jambeck, and Kara Lavender Law. "Production, Use, and Fate of All Plastics Ever Made." *Science Advances* 3, no. 7 (2017): e1700782. doi:10.1126/sciadv.1700782.

Graham, Adam H. "Bans on Plastic Straws Are Growing. But Is the Travel Industry Doing Enough? *New York Times*, May 1, 2018, sec. Travel. https://www.nytimes.com/2018/05/01/travel/straw-bans-hotels-resorts.html.

Gruber, Elisabeth S., Vanessa Stadlbauer, Verena Pichler, Katharina Resch-Fauster, Andrea Todorovic, Thomas C. Meisel, Sibylle Trawoeger, et al. "To Waste or Not to Waste: Questioning Potential Health Risks of Micro- and Nanoplastics with a Focus on Their Ingestion and Potential Carcinogenicity." *Exposure and Health* (2022). https://doi.org/10.1007/s12403-022-00470-8.

Gunnison, Anne, Susan Heald, Jiasun Tsang, Yoonjo Lee, and Jennifer Giaccai. "Preventive Conservation and Identification of Plastic of a Recent Acquisition at the National Museum of the American Indian." In *Object Specialty Group Postprints. American Institute for Conservation 38th Annual Meeting, Milwaukee*, 21–32. Washington, DC: American Institute for Conservation, 2015.

Haagaard, Alex. "Complicating Disability: On the Invisibilization of Chronic Illness throughout History." Platypus, accessed June 8, 2022. https://blog.castac.org/2022/02/complicating-disability-on-the-invisibilization-of-chronic-illness-throughout-history/.

Hammer, Martin, and Christina Lodder. *Constructing Modernity: The Art and Career of Naum Gabo*. New Haven, CT: Yale University Press, 2000.

Haram, Linsey, James T. Carlton, Gregory M. Ruiz, and Nikolai A. Maximenko. "A Plasticene Lexicon." *Marine Pollution Bulletin* 150 (January 1, 2020): 110714. https://www.sciencedirect.com/science/article/pii/S0025326X19308707.

Haraway, Donna J. *Staying with the Trouble: Making Kin in the Chthulucene*. Durham, NC: Duke University Press, 2016.

Hawkins, Gay. "Plastics and Presentism: The Time of Disposability." *Journal of Contemporary Archaeology* 5, no. 1 (2018): 91–102.

Heckman, Heather M. "Burn After Viewing, or, Fire in the Vaults: Nitrate Decomposition and Combustibility." *American Archivist* 73 (Fall–Winter 2010): 483–506.

Hendren, Sara. *What Can a Body Do?: How We Meet the Built World*. New York: Riverhead Books, 2020.

Hicks, Dan. *The Brutish Museums: The Benin Bronzes, Cultural Violence and Cultural Restitution*. London: Pluto Press, 2020.

Jackson, Liz, and Jaipreet Virdi. "Why Won't Nike Use the Word *Disabled* to Promote Its New Go FlyEase Shoe?" *Slate Magazine*, February 5, 2021. https://slate.com

/technology/2021/02/nike-go-fly ease-shoe-disabled-design.html.

Jackson, Zakiyyah Iman. *Becoming Human: Matter and Meaning in an Antiblack World*. New York: NYU Press, 2020.

Jain, Sarah S. "The Prosthetic Imagination: Enabling and Disabling the Prosthesis Trope." *Science, Technology, & Human Values* 24, no. 1 (1999): 31–54.

Jenner, Lauren C., Jeanette M. Rotchell, Robert T. Bennett, Michael Cowen, Vasileios Tentzeris, and Laura R. Sadofsky. "Detection of Microplastics in Human Lung Tissue Using μFTIR Spectroscopy." *Science of the Total Environment* 831 (2022). https://doi.org/10.1016/j.scitotenv.2022.154907.

Joselit, David. "Body Bags." In *Speculations on Anonymous Materials; Nature after Nature; Inhuman*, edited by Susanne Pfeffer, 117–22. London: Koenig Books, 2018.

———. *Feedback: Television against Democracy*. Cambridge, MA: MIT Press, 2007.

———. "Molds and Swarms." In *Part Object Part Sculpture*, edited by Helen Molesworth, 156–65. University Park: Penn State University Press, 2005.

Jue, Melody, and Rafico Ruiz, eds. *Saturation: An Elemental Politics*. Durham, NC: Duke University Press, 2021.

Kafer, Alison. *Feminist, Queer, Crip*. Bloomington: Indiana University Press, 2013.

Katz, Anna. *Eddie Rodolfo Aparicio*. Los Angeles: MOCA, 2023.

Katz, Sylvia. *Plastics and Materials. Designs and Materials*. London: Studio Vista, 1978.

Kline, Cindra. "Santo Domingo Pueblo's Depression Jewelry." *El Palacio*, Spring 2015. https://www.elpalacio.org/2015/03/santo-domingo-pueblos-depression-jewelry.

Kormann, Carolyn. "Where Does All the Plastic Go?" *New Yorker*, September 16, 2019.

Kries, Mateo, Jochen Eisenbrand, and Mea Hoffmann, eds. *Plastic: Remaking Our World*. Weil am Rhein, Germany: Vitra Design Museum, 2022.

Kurzman, Steven. "'There's No Language for This': Communication and Alignment in Contemporary Prosthetics." In *Artificial Parts, Practical Lives: Modern Histories of Prosthetics*, edited by Katherine Ott, David Serlin, and Stephen Mihm, 227–48. New York: NYU Press, 2002.

Le Corbusier. "Type-Needs: Type-Furniture." In *The Theory of Decorative Art: An Anthology of European and American Writings 1750–1950*, edited by Isabelle Frank. New Haven, CT: Yale University Press, 2000.

Leslie, Heather A., Martin J. M. van Velzen, Sicco H. Brandsma, A. Dick Vethaak, Juan J. Garcia-Vallejo, and Marja H. Lamoree. "Discovery and Quantification of Plastic Particle Pollution in Human Blood." *Environment International* 162 (2022). https://doi.org/10.1016/j.envint.2022.107199.

Liboiron, Max. *Pollution Is Colonialism*. Durham NC: Duke University Press, 2021.

Liboiron, Max, and Josh Lepawsky. "There's No Such Thing as We: A Theory of Difference." In *Discard Studies: Wasting, Systems, and Power*, 97–124, Cambridge, MA: MIT Press, 2022.

Mah, Alice. *Plastic Unlimited: How Corporations Are Fuelling the Ecological Crisis and What We Can Do about It*. Hoboken, NJ: John Wiley & Sons, 2022.

Malabou, Catherine. *The Future of Hegel: Plasticity, Temporality and Dialectic*. Translated by Lisabeth During. London: Routledge, 2005. Originally published in 1996 in French as *L'Avenir de Hegel*.

———. *What Should We Do with Our Brain?* Translated by Sebastian Rand. New York: Fordham University Press, 2008.

Maor, Tamar, Angelica Bartoletti, and Bronwyn Ormsby. "Insights into Eva Hesse's Working Practice: A Technical Study of *Addendum*." *Tate Papers* 33 (2020). https://www.tate.org.uk/research/tate-papers/33/eva-hesse-working-practice-technical-study-addendum.

McClure, Ian, Laurence Kanter, and Lisa R. Brody, eds. "Unstable Works of Art." *Yale University Art Gallery Bulletin* (2010): 130–34.

McClure, Ian, and Carol Snow. "Inside Art: Teaching Technical Art

History at the Yale University Art Gallery." *Yale University Art Gallery Bulletin* (2013): 72–81.

McKee, Pat. *Orthotics in Rehabilitation: Splinting the Hand and Body*. Philadelphia: F.A. Davis, 1998.

Meikle, Jeffrey L. *American Plastic: A Cultural History*. New Brunswick, NJ: Rutgers University Press, 1995.

———. "Into the Fourth Kingdom: Representations of Plastic Materials, 1920–1950." *Journal of Design History* 5, no. 3 (1992): 173–82.

———. "Material Doubts: The Consequences of Plastic." *Environmental History* 2, no. 3 (1997): 271–300.

Moore, Charles. *Plastic Ocean: How a Sea Captain's Chance Discovery Launched a Determined Quest to Save the Oceans*. New York: Penguin, 2011.

Nash, Steven, and Jorn Merkert, eds. *Naum Gabo: Sixty Years of Constructivism*. Munich: Prestel-Verlag, 1985.

Nelson, Mallory Kay, Ashley Shew, and Bethany Stevens. "Transmobility: Possibilities in Cyborg (Cripborg) Bodies." *Catalyst: Feminism, Theory, Technoscience* 5, no. 1 (Spring 2019): 1–20.

Nielsen, Tobias D., Jacob Hasselbalch, Karl Holmberg, and Johannes Stripple. "Politics and the Plastic Crisis: A Review throughout the Plastic Life Cycle." *Wiley Interdisciplinary Reviews: Energy and Environment* 9, no. 1 (2020): e360.

Nixon, Rob. *Slow Violence and the Environmentalism of the Poor*. Cambridge, MA: Harvard University Press, 2011.

O'Neill, Kate. *Waste*. Cambridge, UK: Polity, 2019.

Oosten, Thea van. *Properties of Plastics: A Guide for Conservators*. Los Angeles: Getty Conservation Institute, 2022.

Orona, Franke. "Native Americans Urge Deb Haaland to Help Tackle Pollution in Communities of Color." *Hill*, March 25, 2021.

Osodi, George. *Delta Nigeria: The Rape of Paradise*. London: Trolley Books, 2011.

Ott, Katherine. "Disability Things: Material Culture and American Disability History, 1700–2010." In *Disability Histories*, edited by Susan Burch and Michael Rembis, 119–35. Champaign: University of Illinois Press, 2014.

Pagliarino, Amanda V., and Yvonne Shashoua. "Caring for Plastics in Museums, Galleries and Archives." In *Plastics Collecting and Conserving*, edited by Anita Quye and Colin Williamson, 91–98. Edinburgh: National Museums of Scotland, 1999.

Paoletti, Jo Barraclough. *Pink and Blue: Telling the Boys from the Girls in America*. Bloomington: Indiana University Press, 2012.

Price, Beth, Sally Malenka, Ken Sutherland, Andrew Lins, and Janice H. Carlson. "Naum Gabo's Construction in Space: Two Cones: History and Materials." In *Plastics: Looking at the Future and Learning from the Past, Papers from the Conference Held at the Victoria and Albert Museum, London, 23–25 May 2007*, edited by Brenda Keneghan and Louise Egan, 81–88. London: Archetype Publications Ltd., 2008.

Princen, Thomas. "Distancing: Consumption and the Severing of Feedback." In *Confronting Consumption*, edited by Thomas Princen, Michael Maniates, and Ken Conca, 103–32. Cambridge, MA: MIT Press, 2002.

Project Plastics Group. "Plastic Identification Tool." Cultural Heritage Agency of the Netherlands. Accessed February 8, 2023. https://plastic-en.tool.cultureelerfgoed.nl/.

Pullen, Derek. "Managing Change—The Conservation of Plastic Sculptures. Works by Naum Gabo (1890–1977) and Tony Cragg (Born 1949)." In *Material Matters: The Conservation of Modern Sculpture*, edited by Jackie Heuman, 101–7. London: Tate Gallery Publishing Ltd.

Pullen, Derek, and Jackie Heuman. "Cellulose Acetate Deterioration in the Sculptures of Naum Gabo." In *Preprints of Contributions to the Modern Organic Materials Meeting*, 57–66. Edinburgh: SCCR, 1988.

Pullin, Graham. *Design Meets Disability*. Cambridge, MA: MIT Press, 2009.

Rankin, Elizabeth. "A Betrayal of Material: Problems of Conservation in the Constructivist Sculpture of Naum Gabo and Antoine Pevsner." *Leonardo* 21, no. 3 (1988): 285–90.

Rao-Middleton, Ananya. "What I Wish Non-Disabled People Understood about Disability and Plastic." Greenpeace, March 29, 2022. https://www.greenpeace.org.uk/news/disability-ableism-plastic-campaigns/.

Rasmussen, Seth C. "From Parkesine to Celluloid: The Birth of Organic Plastics." *Angewandte Chemie International Edition* 60, no. 15 (2021): 8012–16.

Rathje, William, and Cullen Murphy. *Rubbish! The Archaeology of Garbage*. Tucson: University of Arizona Press, 2001.

Reilly, Julie A. "Celluloid Objects: Their Chemistry and Preservation." *Journal of the American Institute for Conservation* 30, no. 2 (1992): 145–62.

Roberts, Jody A. "Reflections of an Unrepentant Plastiphobe." *Science as Culture* 19, no. 1 (March 2010): 101–20.

Robertson, Kirsty. "Plastiglomerate." *e-flux journal* 78 (December 2016). https://www.e-flux.com/journal/78/82878/plastiglomerate.

Sanderson, Colin, and Christina Lodder. "Catalogue Raisonné of the Constructions and Sculptures of Naum Gabo." In *Naum Gabo: Sixty Years of Constructivism*, edited by Steven Nash and Jorn Merkert, 193–255. Munich: Prestel-Verlag, 1985.

Sangkham, Sarawut, Orasai Faikhaw, Narongsuk Munkong, Pornpun Sakunkoo, Chumlong Arunlertaree, Murthy Chavali, Milad Mousazadeh, and Ananda Tiwari. "A Review on Microplastics and Nanoplastics in the Environment: Their Occurrence, Exposure Routes, Toxic Studies, and Potential Effects on Human Health." *Marine Pollution Bulletin* 181 (2022): 113832.

Selwitz, Charles M. *Cellulose Nitrate in Conservation*. Los Angeles: Getty Conservation Institute, 1988.

Shashoua, Yvonne. *Conservation of Plastics: Materials Science, Degradation and Preservations*. Amsterdam: Elsevier, 2008.

Skeist, Irving, and Jerry Miron. "History of Adhesives." *Journal of Macromolecular Science: Part A—Chemistry* 15, no. 6 (1981): 1151–63, doi:10.1080/00222338108066458.

Smith, Paul Chaat, and Candice Hopkins. *Brian Jungen: Strange Comfort*. Washington, DC: National Museum of the American Indian, 2009.

Smither, Roger, ed. *This Film Is Dangerous: A Celebration of Nitrate Film*. Brussels: Fiaf, 2002.

Sobchack, Vivian. "A Leg to Stand On: Prosthetics, Metaphor, and Materiality." In *The Object Reader*, edited by Fiona Candlin and Rayford Guins, 17–41. New York: Routledge, 2009.

Stewart, Robbie, David Littlejohn, Richard Pethrik, Norman Tennent, and Anita Quye. "The Use of Accelerated Ageing Tests for Studying the Degradation of Cellulose Nitrate." In *ICOM-CC 11th Triennial Meeting Edinburgh 1–6 September, Preprints vol. II*, edited by J. Bridgland, 967–70. London: James & James Ltd.

Stiker, Henri-Jacques. *A History of Disability*. Ann Arbor: University of Michigan Press, 1999.

Summerour, Rebecca, Susan Heald, Sarah Owens, Odile Madden, and Jennifer Giaccai. "Rubberized Flannel in Contemporary Beaded Powwow Regalia." In *Preprints of the 9th North American Textile Conservation Conference, San Francisco, California, November 12–15, 2013*, 183–98.

Tarkanian, Michael J., and Dorothy Hosler. "America's First Polymer Scientists: Rubber Processing, Use and Transport in Mesoamerica," *Latin American Antiquity* 22, no. 4 (2011): 469–86.

Thompson, Richard C., Shanna H. Swan, Charles J. Moore, and Frederick S. Vom Saal. "Our Plastic Age." *Philosophical Transactions of the Royal Society B: Biological Sciences* 364, no. 1526 (2009): 1973–76.

Tolinski, Michael, and Conor P. Carlin. *Plastics and Sustainability Grey Is the New Green: Exploring the Nuances and Complexities of Modern Plastics*. Beverly, MA: Scrivener Publishing, 2012.

Van Sebille, Erik, Chris Wilcox, Laurent Lebreton, Nikolai Maximenko, Britta Denise Hardesty, Jan A. Van Franeker, Marcus Eriksen, et al. "A Global Inventory of Small Floating Plastic Debris." *Environmental Research Letters* 10, no. 12 (2015): 124006.

Wagner-Lawlor, Jennifer A. "The Persistence of Utopia: Plasticity and Difference from Roland Barthes to Catherine Malabou.

Journal of French and Francophone Philosophy 25, no. 2 (2017): 67–86.

———. "Poor Theory and the Art of Plastic Pollution in Nigeria: Relational Aesthetics, Human Ecology, and 'Good Housekeeping.'" *Social Dynamics* 44, no. 2 (2018): 198–220.

Weems, Jason. "War Furniture: Charles and Ray Eames Design for the Wounded Body." *Boom: A Journal of California* 2, no. 1 (2012): 46–48.

Weinhart, Martina, ed. *Plastic World*. Frankfurt: Schirn Kunsthalle, 2023.

Weise, Jillian. "Common Cyborg." In *Disability Visibility: First-Person Stories from the Twenty-First Century*, edited by Alice Wong, 63–74. New York: Vintage Books, 2020.

Wiesenberger, Robert, ed. *Humane Ecology: Eight Positions*. New Haven, CT: Yale University Press, 2023.

Wilson, Kristina. *Mid-Century Modernism and the American Body: Race, Gender, and the Politics of Power in Design*. Princeton, NJ: Princeton University Press, 2021.

Wong, Alice. "The Last Straw." *Eater* (blog), July 19, 2018. https://www.eater.com/2018/7/19/17586742/plastic-straw-ban-disabilities.

Zalasiewicz, Jan, Colin N. Waters, Juliana A. Ivar do Sul, Patricia L. Corcoran, Anthony D. Barnosky, Alejandro Cearreta, Matt Edgeworth, et.al. "The Geological Cycle of Plastics and Their Use as a Stratigraphic Indicator of the Anthropocene." *Anthropocene* 13 (March 2016): 4–17.

Contributors

Kevin Beasley is an artist who lives and works in New York City. His practice spans sculpture, photography, sound, and performance, while centering on materials of cultural and personal significance. Beasley alters, casts, and molds these diverse materials to form a body of works that acknowledge the complex, shared histories of the broader US experience, steeped in generational memories. His work is in museum collections worldwide, including the Guggenheim Museum (New York), Museum of Modern Art (New York), Tate Modern (London), and San Francisco Museum of Modern Art.

Caroline Fowler is Starr Director of the Research and Academic Program at the Clark Art Institute. Her most recent books include *Slavery and the Invention of Dutch Art* (Duke University Press, 2025) and *The Art of Paper: From the Holy Land to the Americas* (Yale University Press, 2019).

Jennifer Gabrys is Chair in Media, Culture, and Environment at the University of Cambridge, where she leads the Planetary Praxis research group. Her work can be found at planetarypraxis.org and jennifergabrys.net.

Anne Gunnison is the Alan J. Dworsky Senior Associate Conservator of Objects at the Yale University Art Gallery. She received a BA in art history from Stanford University, and an MA in principles of conservation as well as an MSc in conservation for archaeology and museums from the Institute of Archaeology at University College London.

Susan Heald has been textile conservator at the Smithsonian's National Museum of the American Indian since 1994. She specializes in preserving textiles and clothing, making them accessible to communities and makers, and working to promote shared stewardship. Heald has authored articles on collaborative preservation efforts with Indigenous communities and artists at the museum.

David Joselit is Arthur Kingsley Porter Professor of Art, Film, and Visual Studies at Harvard University. His most recent books are *Heritage and Debt: Art in Globalization* (MIT Press, 2020) and *Art's Properties* (Princeton University Press, 2023).

Kelly McHugh is the head of conservation at the National Museum of the American Indian, where she has worked since 1996. She plays an active role in the shared stewardship and ethical return of Native American and Indigenous collections. McHugh was a core member in the development of the School for Advanced Research's Guidelines for Collaboration, and serves as a colead of the Smithsonian's Shared Stewardship and Ethical Returns Strategic Priority implementation group.

Elena Torok is the associate objects conservator at the Princeton University Art Museum. She works on the treatment, research, and long-term care of three-dimensional objects in the collections, including archaeological materials, contemporary art, decorative arts, and sculpture. Before joining the museum in 2022, Torok held positions at the Dallas Museum of Art (2017–22) and Yale University Art Gallery (2013–17). She earned her MS from the Winterthur/University of Delaware program in art conservation in 2013 with concentrations in objects conservation and preventative conservation.

Ittai Weinryb is an associate professor at Bard Graduate Center. He is the author of *The Bronze Object in the Middle Ages* (Cambridge University Press, 2016) and *Hildesheimer Avantgarde: Kunst und Kolonialismus in mittelalterlichen Deutschland* (Michael Imhof Verlag, 2023), and the curator of *Agents of Faith: Votive Objects in Time and Place* (Bard Graduate Center, 2018).

Dr. Bess Williamson is an associate professor of design studies at North Carolina State University. Her research focuses on inequities in design, particularly the history of disability access—and lack of access—in modern architecture and industrial design. She is the author of *Accessible America: A History of Disability and Design* (NYU Press, 2019) and coeditor of *Making Disability Modern: Design Histories* (Bloomsbury, 2020).

Index

Photography and Copyright Credits

© 2025 Artists Rights Society (ARS), New York: ADAGP, Paris: 19; SOMAAP, Mexico City. Photo: © Tate: 9

© Arthur Jafa. Courtesy of the artist, Gladstone Gallery, and Sprüth Magers: 15

© The Board of Trustees of the Science Museum: 31

Courtesy Castelli Gallery © 2025 The Andy Warhol Foundation for the Visual Arts, Inc./Licensed by Artists Rights Society (ARS), New York: 21

Courtesy of Chesapeake Bay Program. Photo: Will Parson: 33

Courtesy Clark Art Institute, clarkart.edu. Photo: Tucker Blair: 2

Courtesy of David Liittschwager: 35

Division of Medicine and Science, National Museum of American History, Smithsonian Institution, Gift of Penny Richards and Peter Turley: 30

George Osodi: 22

Courtesy of Duke Riley Studio: 8

Courtesy of the Harvard Art Museums. Photo © President and Fellows of Harvard College: 12

Ifeoma Anyaeji: 23

Courtesy of Kelly Jazvac. Photo: Kelly Wood: 36

© Kevin Beasley. Courtesy Kevin Beasley Studio: 37 (Photo: Kevin Beasley), 38, 39 (Photo: Aundre Larrow)

© Kevin Beasley. Courtesy the artist and Casey Kaplan, New York: 5 (Photo: Ron Amstutz), 6, 40 (Photo: Jason Wyche)

Courtesy of Maphoto/Riccardo Pravettoni: 34

© Matthew Angelo Harrison. Courtesy of the artist and Jessica Silverman, San Francisco. Photo by Timothy Johnson: 16

Digital Image © The Museum of Modern Art/Licensed by SCALA/Art Resource, NY: © The Estate of Eva Hesse. Courtesy Hauser & Wirth: 4; © Succession Marcel Broodthaers c/o Sabam Belgium/ARS, NY 2024: 13, 14; © The Claes Oldenburg Estate: 20; © 2025 Lynda Benglis/Licensed by VAGA at Artists Rights Society (ARS), NY. Gift of the Fuhrman Family Foundation through the Modern Women's Fund: 18

National Archives, NAID: 12168106: 10

National Museum of the American Indian, Smithsonian Institution. Photo by NMAI Photo Services: 26, 27, 28, 29

The Work of Naum Gabo © Nina & Graham Williams/Tate, 2020. Presented to Tate Archive by Nina and Graham Williams, 1993, Photo: Tate: 11; Presented by the artist through the American Federation of Arts 1969, Photo: © Tate: 17; Conservation Department at the Dallas Museum of Art: 25; Image courtesy Dallas Museum of Art: 24

Yale University Art Gallery Gift of Collection Société Anonyme. © 2025 Artists Rights Society (ARS), New York/ADAGP, Paris: 3

ART/WORK

Caroline Fowler and Ittai Weinryb, Series Editors

Books in the ART/WORK series focus on specific media across time periods and cultures, collapsing long-standing asymmetries and bringing the specialist language of conservation to bear on the material history of art. The series presents accessible introductions to art history founded in collaboration, conversation, and an understanding of objects as they were formed through making, deterioration, care, and remaking.

Plastics
Anne Gunnison, David Joselit

Pigments
Barbara H. Berrie, Caroline Fowler, Karin Leonhard, and Ittai Weinryb

Ceramic Art
Margaret S. Graves, Sequoia Miller, Magdalene Odundo, and Vicki Parry